Solar Inverter Design with Improved Performance

By
Mona Reyes

3.4 SYMMETRIC MODULAR MULTILEVEL
INVERTER TOPOLOGY 70

3.5 COMPARISON OF THE PROPOSED
9-LEVEL SYMMETRIC MMI WITH
CONVENTIONAL 9-LEVEL INVERTER
TOPOLOGIES. 74

3.6 ASYMMETRIC MODULAR MULTILEVEL
INVERTER TOPOLOGY 76

3.7 COMPARISON OF THE PROPOSED
27-LEVEL ASYMMETRIC MMI WITH
CONVENTIONAL 27-LEVEL INVERTER
TOPOLOGIES 86

3.8 ASYMMETRIC 31-LEVEL MODULAR
MULTILEVEL INVERTER TOPOLOGY 89

3.9 POWER LOSSES CALCULATION 97

3.9.1 Conduction Losses 97

3.9.2 Switching Losses 99

3.9.3 Loss Calculation of the Proposed
27-level MMI 101

3.10 SELECTIVE HARMONIC PULSE WIDTH MODULATION FOR THE PROPOSED MMI   103

3.11 SIMULATION RESULTS   107

   3.11.1 9-Level MMI   107

   3.11.2 11-Level MMI   110

   3.11.3 13-Level MMI   112

   3.11.4 15-Level MMI   113

   3.11.5 17-Level MMI   115

3.11.6 19-Level MMI 117

3.11.7 21-Level MMI 119

3.11.8 23-Level MMI 122

3.11.9 25-Level MMI 124

3.11.10 27-Level MMI 126

3.11.11 29-Level MMI 128

3.11.12 31-Level MMI 130

3.12 EXPERIMENTAL RESULTS 132
3.12.1 9-level Asymmetric Modular Multilevel Inverter 133

3.12.2 27-level Asymmetric Modular Multilevel Inverter 136

3.12.3 31-level Asymmetric Modular Multilevel Inverter 136

3.13 SUMMARY 138

HARMONIC ELIMINATION IN MODULAR MULTILEVEL INVERTER USING OPTIMIZATION TECHNIQUES

139

4.1 INTRODUCTION — 139

4.2 PROBLEM STATEMENT — 141

4.3 OPTIMIZATION TECHNIQUES — 147

4.4 BEES ALGORITHM — 150

4.4.1 Natural World of Bees — 150

4.4.2 Computation of Switching Angles — 151

4.5 GENETIC ALGORITHM — 154

PARTICLE SWARM OPTIMIZATION 157

SIMULATION RESULTS 160

4.7.1 Implementation of SHE for
9-Level MMI 160

4.7.1.1 BEE colony optimization   162

4.7.1.2 Genetic algorithm   164

4.7.1.3 Particle swarm optimization   167

4.7.2 Implementation of SHE for

27-Level MMI   170

4.8   EXPERIMENTAL RESULTS   179

4.9   SUMMARY   184

5 DESIGN AND IMPLEMENTATION OF MODULAR MULTILEVEL INVERTER FOR GRID-CONNECTED PV SYSTEM 185

5.1	INTRODUCTION	185
5.2	PROBLEM STATEMENT	189
5.3	DESIGN AND IMPLEMENTATION OF MMI BASED GRID CONNECTED PV SYSTEM	191
5.4	MODELING OF PV SYSTEM	192
5.5	DESIGN OF DC-DC BOOST CONVERTER FOR MMI	195
	5.5.1 Design of Boost Converter for Symmetrical 9-Level MMI	198
	5.5.2 Design of Boost Converter for Asymmetrical 31-Level MMI	199

10

5.6	MPPT OF PV SYSTEM USING BOOST CONVERTER	199
	5.6.1 Perturb and Observe Algorithm	201
5.7	DESIGN OF MODULAR MULTILEVEL INVERTER	203
	5.7.1 DC Input of Modular Multilevel Inverter	204
	5.7.2 Relationship Between DC Input Power and AC Output Power of MMI	204
5.8	DESIGN OF LCL FILTER FOR GRID CONNECTED SYSTEM	206
5.9	CONTROL OF THE SUGGESTED GRID-TIED SYSTEM	208
	5.9.1 Total DC link voltage Controller	209
	5.9.2 Control of Grid Current and Grid Voltage Tracking	210
	5.9.3 Control of Independent DC Link Voltages	212

5.10 PI CONTROLLER   213

5.11 TUNING OF CONTROLLER   214

5.12 TUNING USING OPTIMIZATION

ALGORITHMS   215

5.12.1 Particle Swarm Optimization   216

5.12.2 Firefly Algorithm   220

5.12.2.1 Light intensity and

attractiveness   220

5.12.2.2 Distance   222

5.12.2.3 Location update   222

5.12.3 Harris Hawks Optimization (HHO) 223
RESULT AND DISCUSSION 231
Implementation of PSO-PI Controller 233
Implementation of FFA-PI Controller 235
Implementation of HHO-PI Controller 237
Experimental Validation 241
5.6 SUMMARY 246

6 SUMMARY AND SUGGESTIONS FOR FURTHER RESEARCH 247

SUMMARY 247
SUGGESTIONS FOR FURTHER RESEARCH 250

LIST OF TABLES

TABLE NO.	TITLE	PAGE NO.
1.1	Equations of the Proposed Symmetric and Asymmetric Modular Multilevel Inverter	26
2.1	Literature Review of MMI	59
3.1	Switching Sequence of the Symmetric MMI	73
3.2	Parameters of the Symmetric MMI	73
3.3	Current Path of 9-Level MMI	74
3.4	Comparison of the suggested Symmetric MMI with conventional 9-Level Symmetrical Topologies	75
3.5	Firing Sequence of the 27-Level AMMI	78
3.6	Parameters of the 27-Level AMMI	79
3.7	Current Path of 27 -Level MMI	82
3.8	Possible Levels generated by the AMMI	84
3.9	Comparison of 27-Level Inverter Topologies	87
3.10	Switching States of the Proposed 31-Level Inverter	89
3.11	Parameters of the 31-Level MMI	91
3.12	Current Path of 31-Level MMI	94
3.13	Comparison of The Proposed 31-Level Inverter with conventional 31-Level Inverter Structures	96
3.14	Simulation Parameters	108
3.15	Triggering Sequence for 11-Level AMMI	110
3.16	Triggering Sequence for 13-level AMMI	112
3.17	Triggering Sequence for 15-Level AMMI	114

Triggering sequence for 17-Level AMMI 116

Triggering sequence for 19-Level AMMI 118

Triggering sequence for 21-Level AMMI 120

Triggering sequence for 23-Level AMMI 122

Triggering sequence for 25-Level AMMI 124

Triggering sequence for 29-Level AMMI 128

Comparison of Results for the Proposed MMI 137

BCO Parameters 162

Triggering Angles for different Modulation

Index and THD for BCO 162

GA Parameters 165

Triggering Angles for different Modulation

Index and THD for GA 165

PSO Parameters 167

Triggering Angles for different Modulation

Index and THD for PSO 167

Performance comparison of THD and Harmonic Components
of 9-level MMI with BCO,GA and

PSO 169

Switching Angles for 27-level MMI 173

Comparison of Results for the Proposed

Methodologies 183

Comparison with Other Modulation Methods 183

Parameters of Boost Converter 199

Functioning of P &O Algorithm 203

Specifications of PV Fed Grid Connected

System 231

Parameters of PSO 234
Parameters of FFA 235
Parameters of HHO 237
Gain Parameters and THD Value of the
Proposed Grid Connected system 241
Comparison of Results with Other
Methodologies 245
Comparison of Results for a Solar Fed MMI 249
Comparison of Results with Existing Inverter
Topologies 249

LIST OF FIGURES

FIGURE NO. TITLE PAGE NO.

Grid Connected PV Topologies 5
Full Bridge Inverter with its Output Waveform 8
(a)Waveform of 3-level Inverter
(b)Waveform of 9-level Inverter 12
Circuit diagram of 9-level CHBMLI 15
Circuit diagram of 9-level DCMLI 17
Circuit diagram of 9-level FCMLI 19
Proposed Solar PV Fed Modular Multilevel Inverter 25
Nine-level output voltage waveform 27
2.1 Survey on MMI with Modulation Methods 39
(a) Structure of Diode Bridge Bidirectional switch
(b) Structure of Common Emitter switch
(c) Structure of common Collector switch
(d) Structure of Reverse Blocking Bidirectional switch 66
(a) Basic structure of the MMI (b) Basic unit of MMI. 68
Cascaded structure of MMI 69
Possible voltage levels obtained using symmetric
and asymmetric DC sources 70
Modes of Operation of Symmetric 9-level MMI 71
9-level output of the proposed Symmetric MMI 72
Modes of operation of Asymmetric 27 -level MMI 81
Switching Operation of Asymmetric 27 -level MMI 81
Modes of Operation of Asymmetric 31-level MMI 93

(a) Comparison of Switching Losses of 27-level AMMI with conventional 27- Level Configurations (b)Comparison of Conduction Losses of 27-level AMMI with conventional 27- Level Configurations

(c) Comparison of Total loss of 27-level AMMI with conventional 27- Level Configurations 103

(a)9-Level output Voltage (b) FFT Analysis

of 9-Level Inverter 109

(a)11-Level Output Voltage (b) FFT Analysis

of 11-Level Inverter 111

(a)13-Level Output Voltage (b) FFT Analysis

of 13-Level Inverter 113

(a) 15-Level Output Voltage (b) FFT Analysis

of 15-Level Inverter 115

(a) 17-Level output Voltage (b) FFT Analysis

of 17-Level Inverter 117

(a) 19-Level Output Voltage (b) FFT Analysis

of 19-Level Inverter 119

(a) 21-Level Output Voltage (b) FFT Analysis

of 21-Level Inverter 121

(a) 23-Level Output Voltage (b) FFT Analysis

of 23-Level Inverter 123

(a) 25-Level Output Voltage (b) FFT Analysis

of 25-Level Inverter 126

(a) 27-Level Output Voltage (b) FFT Analysis

of 27-Level Inverter 127

(a) 29-Level Output Voltage (b) FFT Analysis

of 29-Level Inverter 130

(a) 31-Level Output Voltage (b) FFT Analysis
of 31-Level Inverter 131
Block Diagram Representation of Hardware Implementation 132
1kWp Solar PV Plant 133
Prototype of the Proposed MMI 133
(a) Output Voltage of 9- Level MMI

a. Output current of 9- Level MMI
b. Voltage THD of 9-Level MMI
c. Current THD of 9-Level MMI 135

(a) Output Voltage of 27- Level MMI

a. Output current of 27- Level MMI
b. Voltage THD of 27-Level MMI
c. Current THD of 27-Level MMI 137

(a) Output Voltage of 31- Level MMI

a. Output current of 31- Level MMI
b. Voltage THD of 31-Level MMI
c. Current THD of 31-Level MMI 138

Output of 9-level MMI 142

(a) Switching Pulses for unidirectional Switches

of 9-Level MMI (b) Switching Pulses for Bidirectional Switches
of 9-Level MMI (c) Switching Pulses for

P1 and P2 Switches of 9-Level MMI 161

Modulation Indices Vs Optimum Value of Cost

Function using BCO 163

(a) 9-Level output voltage Using BCO

a.   FFT analysis of 9-Level MMI Using BCO 164

Modulation Indices Vs Optimum Value of

Fitness Function Using GA 165

(a) 9-Level output voltage Using GA (b) FFT

analysis of 9-Level MMI Using GA 166

Modulation Indices Vs Optimum Value of

Fitness Function Using PSO 168

(a) 9-Level output voltage Using PSO (b) FFT

analysis of 9-Level MMI Using PSO 168

(a) Switching Pulses for unidirectional Switches of 27-Level MMI
(b) Switching Pulses for Bidirectional Switches of 27-Level MMI

a.   Switching Pulses for P1 and P2 Switches

of 27-Level MMI 172
Modulation Index Vs Switching Angles
for 27-Level MMI 173
(a)Best fitness value for BCO (b) Best fitness value
for GA (c) Best fitness value for PSO 175
(a) 27-Level output voltage Using BCO
(b) FFT analysis of 27-Level MMI Using BCO 176
(a) 27-Level output voltage Using GA
(b) FFT analysis of 27-Level MMI Using GA 177
(a) 27-Level output voltage Using PSO
(b) FFT analysis of 27-Level MMI Using PSO 178
Solar Plant of rating 1kWp 179
Hardware set up of MMI with SHE 180
(a) Experimental output of the 9-level MMI
(b) FFT analysis of the 9-level MMI 181

(a) Experimental output of the 27-level MMI

a.   FFT analysis of the 27-level MMI 182

(a)Single-stage PV system configuration (b)Two-stage PV system configuration

a.   Multi-stage PV system configuration 187

Power flow in single Phase PV system for unity
power factor 188
General Structure of MMI Based Single-Phase
Grid Connected Solar-PV System. 191
Equivalent circuit of the PV cell 193
PV panel characteristics under shadowed condition 195
Solar PV Fed Boost Converter 196
Solar PV Fed MMI 197
Solar-PV array with P&O MPPT controller 200
__Change in $\frac{dP}{dv}$ value of the PV module 201
Flowchart of the P&O algorithm 202
Closed-loop Control system of the proposed GCPVS 211
General structure of a single-phase PLL 212
PI controller 213
Flowchart of PSO tuned PI controller 219
Flowchart of FF tuned PI controller 223
HHO algorithm 225
Flowchart of The HHO Tuned PI Controller 230
Input PV Sources For 31-Level AMMI 232
Total Dc Link Voltage 232
Output voltage and current of 31-level MMI 233

(a) Simulation Result of Grid Voltage Using PSO-PI Controller (b) Simulation Result

of Grid Current Using PSO-PI Controller

(c) FFT Analysis of Grid Current Using

PSO-PI Controller 234

(a) Simulation Result of Grid Voltage Using FFA-PI Controller (b) Simulation Result of Grid Current Using FFA-PI Controller (c) FFT Analysis of

Grid Current Using FFA-PI Controller 236

Output Waveforms of Grid Voltage, Grid Current,

Active and Reactive Power Using HHO-PI controller 238

FFT analysis of the grid current using HHO 238

Output Waveforms of Grid Voltage, Grid Current,

Active and Reactive Power during transient

condition Using HHO-PI controller 239

Performance of HHO-PI controller during

Voltage Sag 240

Performance of HHO-PI controller during

Voltage Swell 240

Prototype model of the suggested system 242

(a) Voltage waveform of the 31-Level MMI(experimental) (b) Current waveform of

the 31-Level MMI(experimental) 243

Grid Voltage and Grid Current of the Prototype

Under Steady State Condition 244

Experimental THD of Grid Current 244

Experimental Output of Grid Voltage and Grid

Current Under Transient Condition 245

LIST OF SYMBOLS AND ABBREVIATIONS

- Angular frequency

y - Absorption Coefficient

AC - Alternate Current

ANN - Artificial Neural Network

ASMLI - Asymmetric Multilevel Inverter

AMMI - Asymmetrical Modular Multilevel Inverter

β_f - Attractiveness

BCO - Bee Colony Optimization

CMLI - Cascade Multilevel Inverter

CHB - Cascaded H-Bridge

CHBMLI - Cascaded H-Bridge Multi Level Inverter

- Change in time

CC - Common Collector

CE - Common Emitter

CM - Common Mode

CMOS - Complementary metal–oxide–semiconductor

p_{cdn} - Conduction loss

- Conversion efficiency

CSS-MLI - Cross Connected Source-Multi Level Inverter DDR - Device Duty Ratio

DSO - Digital Storage Oscilloscope

DB - Diode Bridge

DCMLI - Diode Clamped Multilevel Inverter

DC - Direct Current
DFIG - Double Fed Induction Generator
- Efficiency
EMC - Electro Magnetic Compatibility
ELPSO - Enhanced Leader Particle Swarm Optimization
FFT - Fast Fourier Transform
FPGA - Field Programmable Gate Array
FFA - Firefly Algorithm
FSO - Fish Swarm Optimization
FC - Flying Capacitor
FCMLI - Flying Capacitor Multilevel Inverter
FSFPWM - Fundamental Switching Frequency Pulse width

Modulation

GTO - Gate Turn-Off
GA - Genetic Algorithm
GMPPT - Global Maximum Power Point
GCPV - Grid Controlled Photovoltaic
GVPVS - Grid Controlled Photovoltaic Systems
GCC - Grid Current Controller
GCS - Grid-Connected System
HDL - Hardware Description Language
HHO - Harris Hawks optimization
HAPF - Hybrid Active Power Filter
InC - Incremental Conductance
IGBT - Insulated Gate Bipolar Transistor
IAE - Integral Absolute Error
Ki - Integral gain

ISE - Integral Square Error

ITAE - Integral Time Absolute Error

ITSE - Integral Time Square Error

LF - Levy Flight

MPP - Maximum Power Point

MPPT - Maximum Power Point Tracking

MOSFET - Metal Oxide Field Effect Transistor

MPC - Modern Predictive Controller

MMI - Modular Multilevel Inverter

MTO - MOS Turn-Off

MLI - Multi Level Inverter

MCPWM - Multicarrier Pulse Width Modulation MDSCF - Multiple Delayed Signal Cancellation Filter NLC - Nearest Level Control

NLC-PWM - Nearest Level Control-Pulse Width Modulation NPC - Neutral Point Clamped

NPCMLI - Neutral Point Clamped Multi Level Inverter

N_L - Number of levels

N_{LC} - Number of levels obtained by cascaded connection

PSO - Particle Swarm Optimization

PIV - Peak Inverse Voltage

P&O - Perturb and Observe

PLL - Phase Locked Loop

PV - Photovoltaic

PLECS - Piecewise Linear Electrical Circuit Simulation

PCC - Point of Common Coupling

PF - Power Factor

PI - Proportional and Integral

Kp - Proportional gain

PID - Proportional Integrative Derivative

PR - Proportional Resonant

PWM - Pulse Width Modulation

PD-PWM - Pulse Width-Pulse Width Modulation

α - Randomization factor

RES - Renewable Energy System

RG-IGBT - Reverse blocking-IGBT

RMS - Root Mean Square

SHE - Selective Harmonic Elimination Sic-MOSFETs - Silicon Carbide MOSFETs

SPWM - Sinusoidal Pulse Width Modulation

SMC - Sliding Mode Control

SVPWM - Space Vector Pulse Width Modulator

STC - Standard Test Condition

SoC - State of Charge

SVM - Support Vector Machine

SCMLI - Switched Capacitor Multilevel Inverter

SDDS - Switched Diode Dual Source

SRM - Switched Reluctance Motor

β - Switching angle

p_{sw} - Switching loss

SMLI - Symmetric Multilevel Inverter

SMMI - Symmetrical Modular Multilevel Inverter

TLBO - Teaching-Learning-Based Optimization

THIPWM - Third Harmonic Injection Pulse Width Modulation

THD - Total Harmonic Distortion
TSV - Total Standing Voltage
VLB - Voltage Level Boost
Ψ - Voltage ripple
VSI - Voltage Source Inverter
ZA - Zero Oscillation Algorithm
Z-N - Ziegler–Nichols

CHAPTER 1 INTRODUCTION

ROLE OF RENEWABLE ENERGY

The demand for electricity is steadily increasing due to the increase in inhabitants and industrial development. Energy generated from natural sources such as coal or gas is not consistent with today's energy demand. Since coal (fossil fuel) is cheaply available, an enormous quantity of electrical energy is generated from fossil fuels. Fossil fuels cause significant damage to the environment by liberating carbon dioxide and mercury during energy conversion, which leads to global warming Gulagi *et al.* (2020). Sustainable energy sources can be used in power generation to overcome the challenges faced by generating electrical power from fossil fuels (Stonier & Lehman 2018).

Abdelrazik *et al.* (2018) dealt with the net generation of electricity from renewable sources which is expected to be equal to coal production by 2040. Nearly half of renewable energy is obtained from wind and solar power. Recent advances lead to renewable energy systems which are low cost. The average cost reduction for Solar Photovoltaic (PV) and shore winds forecast at 40–70% and 10-20 % by 2040. The growing global energy crisis is affected by the depletion of conventional energy sources (Albert & Stonier 2020), and also has the following drawbacks:

◇ Emission of harmful pollutants

◇ Carbon dioxide and mercury emissions are anticipated to increase by 35% and 8% by 2020 as the electricity generation is projected to increase

Global warming is estimated to raise the earth's surface temperature from 3^0C to 6^0C by the end of this century as a result of the green house effect. The power generated by conventional systems needs to be passed on to

the end-user on a long-term basic basis, requiring costly, complex infrastructure and exposing the entire system to higher energy loss and safety risks.

Kumar *et al.* (2019) encouraged investments in clean-energy production through PV systems. Modern topologies of power converters have provided various positive approaches for the relation to Photovoltaic (PV) systems. In view of fossil fuel price inflation and diminishing public acceptance of these energy sources, photovoltaic technology has become a truly sensible alternative. The PV has the advantage of converting sunlight directly into electricity and is also well suited for most geographical regions. Hence it is highly preferred compared to other Renewable Energy Systems (RES). In particular, solar power has been pollution-free, easily accessible from sunlight, and at a more acceptable level for the past two decades in terms of a minimum price.

There are many types of RES available but the type of source to choose depends on location, usability, and availability. Some of the primitives are as follows:

⬧ Energy demand and significant energy alternatives lead to investment in photovoltaic power generation among all energy sources since energy can be produced between a few kW and several MW depending on geography.

⬧ Most of the regions in the Indian subcontinent are blessed with daily global radiation of 4 -7 kWh/m2/day.

The solar power system is usually the best choice for most suburban and rural applications as it requires less maintenance, provides noise-less operations due to no moving parts, and occupies less space as rooftops can be used to install solar photovoltaic plates. The use of PV systems in energy generation began in the 1970s and today, despite the high capital costs referred to by Khalid (2020), it is developing rapidly worldwide.

PV SYSTEMS IN ENERGY CONVERSION

PV system performance depends largely on the efficiency of solar radiation, temperature, and conversion. Although PV systems have many disadvantages, due to weather variations they suffer from varying system

performance, high installation costs, and low module efficiency up to 20 %.

Optimizing the sizing of stand-alone or grid-connected PV systems (GCPVS) is a convoluted problem of optimization that anticipates obtaining appropriate energy and economic costs for consumers.

Solar energy has the benefits of being employed almost in all places with the proper positioning of PV panels. In general, the advantages of Solar PV systems are:

◇ Low environmental effects

◇ easily mounted close to the consumer thereby reducing the transmission line losses

◇ reduces maintenance costs

◇ environment friendly, since there is zero emission of carbon dioxide gasses

Basically, there exists two categories of PV systems such as (1) standalone and (2) grid connected. In a standalone PV system, additional batteries are required to store the excess amount of produced power. But in a GCPVS the grid functions as a battery with an infinite storing capacity. It takes care of fluctuations in seasonal load. The excess amount of electrical energy produced by the PV system can be stored in the grid without throwing away. Hence, the efficiency of a GCPVS will be higher than the stand-alone PV system. Therefore, the usage of grid PV systems is much desired over standalone PV systems Sabry *et al.* (2020).

The PV installations are increasing rapidly, with novel control strategies for suitable operation. The output of RES can be controlled with the benefits of advanced power electronic converters. The power electronic devices are interfaced with RES for reducing cost, increasing reliability, high efficiency, and power stability. Numerous topologies are designed for GCPVS. Mostly five different structures are used for GCPV power applications.

Figure 1.1 shows the five different inverter configurations for GCPVS. They are the centralized inverter system, the multi-central inverter system, the string inverter system, the multi-string inverter system, and the modular inverter system Kabalci (2020). Figure 1.1a shows the structure of the centralized inverter system. It consists of a single inverter with several strings of PV panels connected in series. The Central inverter pumps AC power to the grid by converting the DC power obtained from the PV panels. The major drawback of centralized inverter system is the absence of maximum power point operation of every solar module during the shading conditions. Figure 1.1b shows the structure of the multi-centralized inverter system in which DC-DC converters are implemented to overcome the drawbacks of central inverter system.

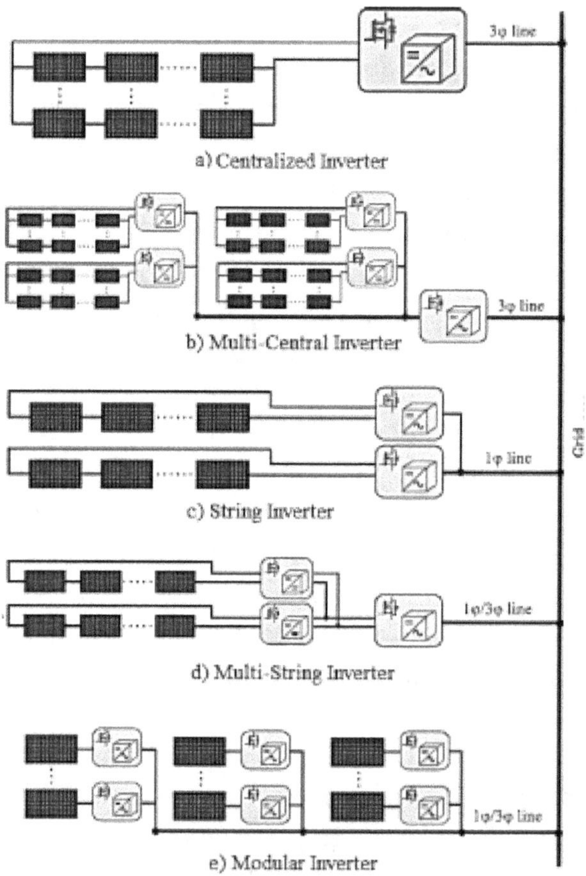

Figure 1.1 Grid Connected PV Topologies

Figure 1.1c shows the structure of a string inverter. The structure is built by connecting PV string to the inverter. A PV string consist of series connected PV modules. In general, the output of the string inverter will range from 340V to 510V.Figure 1.1d shows the structure of a multi-string inverter system. If DC-DC boost converters are connected to the string, then some of the PV panels can be removed from the string. This structure saves cost and is more reliable.

Figure 1.1e shows the structure of a modular inverter . This is also known as micro inverter or module integrated inverter. Because of the compact design of the inverter it can be easily fixed behind the PV module. The bulk manufacturing of these inverters may lower the production cost, and therefore decreases the cost of inverter per watt power generation. The major advantage of this system is that it is free from mismatch losses and works with highest (Maximum Power Point Tracking) MPPT accuracy. The proposed research work uses this type of system for grid integration.

POWER ELECTRONIC CONVERTERS IN RENEWABLE ENERGY CONVERSION

Researchers have focused on renewable energy resources for decades, and numerous power inverters are built to interconnect these technologies into the distribution grid. High voltage power electronic circuits are required in the transmission lines to ensure the delivery of power and power quality. Therefore, power electronic inverters are responsible for performing these conversion tasks efficiently.

The growth of global energy demand has led to the emergence of novel topologies of semiconductors and power converters which are capable of supplying all the required power. There is still an ongoing competition to produce semiconductors which can withstand higher voltage and current for efficient systems. Moreover, there has been heavy competition between the use of high-voltage semiconductor devices in traditional power converter topologies with medium voltage devices in modern converter topologies.

In industries power inverters are used in electrical drive systems. These are generally beneficial for many applications such as transportation (train traction, ship propulsion, and car applications), energy conversion, processing, mining, and petrochemical applications. Many of these processes have steadily increased demand for power to achieve greater manufacturing cost, cost reduction.

The researchers in the domain of power electronics have responded to this demand in two ways:

◈ The development of semiconductor technologies that achieve greater nominal voltages and currents (at present 8 kV and

6 kA), retaining conventional converter configurations (mainly

two-level and current source inverters)

◈ Emerging novel configurations using conventional power semiconductor technology.

The first approach gave birth to popular circuit structures and control techniques. In addition, the novel semiconductor devices are costlier, and additional power filters are required to fulfil the power quality requirements. It is therefore easy to choose to construct a new structure of converter based on a multilevel concept. Right now, this is a challenging issue.

CONCEPT OF CLASSICAL INVERTER

There is a heavy competition currently between the use of conventional configurations implemented with high voltage rating switches and modern converter configurations implemented with medium voltage rating switches. Figure 1.2 illustrates a full bridge inverter. The configuration of full bridge inverter comprises of four switches (S_1, S_2, S_3, S_4) with their respective diodes and one separate voltage source, V_z. When the switches S_1 and S_2 are triggered, the load voltage obtained is $+V_z$. When the switches S_3 and S_4 are triggered the load voltage obtained is $-V_z$.

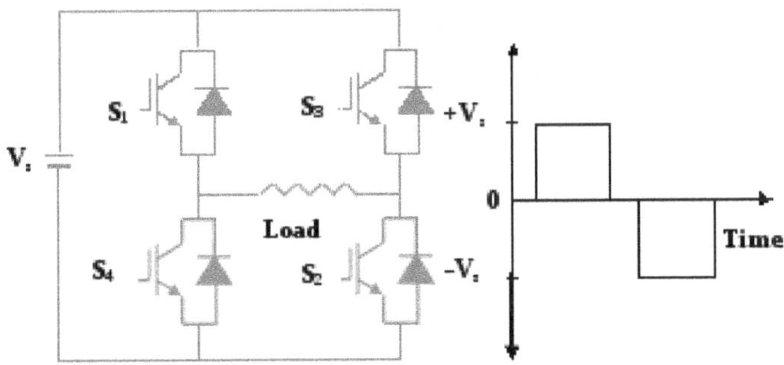

Figure 1.2 Full Bridge Inverter with its Output Waveform

A square waveform is obtained as the output from the positive and negative half cycles which are specified as two levels. The zero potential is also obtained which is also included as an additional level. Therefore, the full

bridge inverter produced three levels of output voltage ($+V_z$, $0V_z$, $-V_z$).

Conduction state of a switch is defined as the state in which the semiconductor switch is triggered. Likewise, the blocking state is defined as the state in which the semiconductor switch is turned off. These two states determine the current and voltage rating of the semiconductor switch. The maximum current carried by a semiconductor switch during the conduction state determines the current rating of the semiconductor switch.

During the blocking state the semiconductor switch has to withstand a voltage. This voltage is called blocking voltage or standing voltage for the semiconductor switch. The maximum voltage blocked by a semiconductor switch during the blocking state determines the voltage rating of a semiconductor switch. Moreover, in a full bridge inverter, the voltage stress is equal to the operating voltage of the inverter. The operating voltage is the peak value of the square wave output.

The Total Standing Voltage (TSV) of the full bridge inverter is obtained by adding the blocking voltages of individual semiconductor switches.

$$TSV = V_z + V_z + V_z + V_z = 4V_z \quad (1.1)$$

The per unit representation of TSV is presented in Equation 1.2.

$$TSV_{p.u.} = 4\,V \quad (1.2)$$

The output voltage of the full bridge inverter $(+V_z, -V_z)$

periodically changes with time. This change in voltage has a serious impact on the load. If a motor is taken as the load, then the change in voltage causes leakage of current in the insulation of the motor. Therefore, the conductor and insulator arrangement inside the motor behaves like a capacitor and the current through the capacitor depends on the rate of change of voltage across it.

$$i(t) = C \frac{dV_c(t)}{dt} \quad (1.3)$$

load.

This rate of change of voltage is determined as the $\frac{dv}{dt}$ stress on the

$$\frac{dv}{dt} \text{ stress on the load} = \lim_{\Delta t \to 0} \frac{\text{change in voltage in a step}}{\Delta t \cdot \text{operating voltage}} \quad (1.4)$$

For the full bridge inverter, the $\frac{dv}{dt}$ stress on the load is

$$\frac{dv}{dt} \text{ stress in p.u} = \lim_{dt \to 0} 2 \quad (1.5)$$

Moreover, with traditional configuration, this converter has the capability to produce only three levels at the output. One of the utmost drawbacks of the three level inverter is the quality of the output voltage (Mukundan *et al.* 2020). This three-level output typically consists of 3_{rd}, 5_{th}, 7_{th}, lower-order harmonics.

These harmonic contents significantly affect the reliability of the equipment. Low-pass filters are being used at the output end of three-level inverter to remove the harmonics. If the quality of power is poor at the output end, then a larger size of filter is implemented. Moreover, it has been well known that even the design of low pass filter is a difficult job and mostly its size is bulky. Numerous works have been conducted by the researchers to reduce the filter size.'

Overall, some of the issues with traditional inverters can be summarized as follows:

◈ During the switch off condition in a three-level inverter, each switch possess the entire DC voltage. This voltage is greater than the voltage of every switching device.

◈ Static voltage sharing is not possible in a three-level inverter since different leakage currents are produced during the turn off condition. To implement static sharing parallel resistors can be used.

◈ Dynamic voltage sharing is not possible during switching, due to the changes in switching frequency. For dynamic voltage sharing specific gate drive methods and snubbers are essential.

◈ Insulation failure occurs in motors run by three-level inverters because of the large step on the load.

◈ Three-level inverters produce more harmonics.

◈ Three-level inverters experience high thermal stress, mechanical stress and ultrasonic vibration induced by inductive elements.

◈ Three-level inverters experience high voltage cable issues, high power losses, high maintenance, high power interference, and poor reliability.

◈ Additional cooling system is also required for three-level inverters.

In the past, three-level inverters have been suitable for medium and high voltage applications. In today's scenario, MLIs utilizing medium-voltage based semiconductors are competing with traditional inverters in the fields of utilization, sophisticated control, advanced semiconductor switches, and in production. While traditional inverters seek their application in low-power systems, they do not satisfy high-power demands. MLIs are the best alternative for energy conversion in medium voltage high power applications (Sathik *et al.* 2020).

BASIC CONCEPT OF MULTILEVEL INVERTER

MLI comprises of a set of power electronic switches and DC sources. Turning on the semiconductor switches adds up a DC voltage to provide a high voltage at the output. During the conversion process the power electronic switches of MLI experience high voltage stress. The three-level inverter generates a two-valued output voltage with a zero potential.

A common structure of three-level, and nine-level inverter with its corresponding waveforms are shown in Figure 1.3. For all such cases, the switching elements are not organized in a sequence but they are organized in such a manner that it can produce three, and nine levels of output voltages. Here, an important thing should be noted that as the steps in the output staircase waveform increases the harmonic distortions tend to decrease. This would greatly improve the power quality of the inverters output.

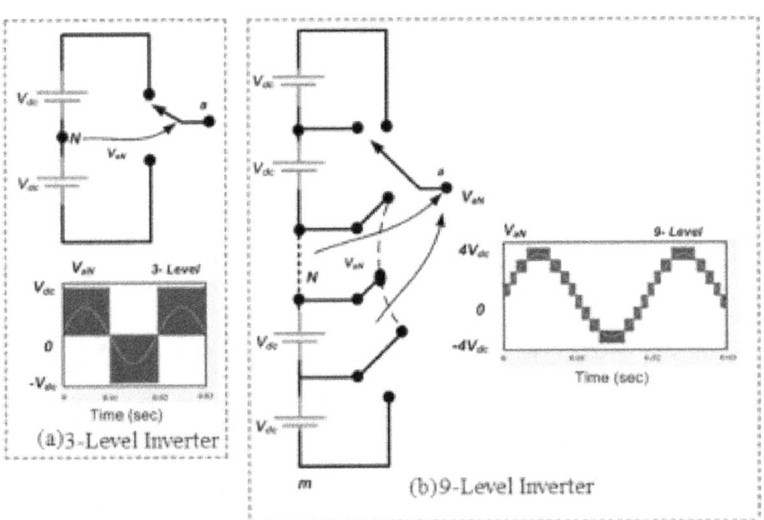

Figure 1.3 (a) 3-level Inverter (b) 9-level Inverter

However, various multilevel structures are effectively designed and tested to generate a staircase wave shape in the output. Therefore, the general concept of multilevel inverters is that the

AC voltage is synthesized at the output from various DC voltages. As the DC sources in the input side tend to increase, the output voltage waveform adds further steps and looks closer to a sinusoidal waveform.

Merits of Multilevel Inverter

In general, a MLI has several advantages when compared with a standard three-level inverter. The striking features are:

_1. Quality of waveform: MLI will not only generate a less harmonic output voltage but also lowers the $\frac{dv}{dt}$ stress. Therefore, the issues related to Electromagnetic Compatibility (EMC) problems can be reduced

2. Common-Mode (CM) voltage: These inverters produce lower CM voltage so that the bearing stress in a motor associated with the multilevel drive is lessened. In addition, the CM voltage can be removed using improved modulation methods

3. Input current: MLI can generate distortion less input current

4. Switching frequency: The MLI can be controlled using fundamental frequency and high frequency switching modulation methods. When MLIs operate using Pulse Width Modulation (PWM) with lower switching frequency, the switching losses are greatly reduced and greater efficiency can be achieved.

Multilevel inverters also have drawbacks. One specific drawback is that MLI requires a large amount of power electronic switches. Even though switches of a lower rating are used in a MLI, a specific gate drive circuit is required for each switch. This makes the entire system high-priced and complicated. However, major advantages over traditional three-level inverters overweigh the advantages of MLIs.

TRADITIONAL TOPOLOGIES OF MULTILEVEL INVERTER

In general, there are three basic types of multilevel inverters namely Neutral Point Clamped (NPC), Flying Capacitor (FC), and Cascaded H-Bridge (CHB) Haghdar (2020).

Cascaded H-Bridge Multilevel Inverter

The CHB inverter is the first semiconductor-based MLI and was developed by Baker and Banister in 1975. This topology contains a series interface of single-phase H-bridge inverters. The requirement of clamping diodes or voltage capacitors is absent in this topology. This configuration uses separate DC sources. Batteries, fuel cells, and solar cells might be used as DC sources. Each H-bridge consists of four switches with their respective diodes and one separate voltage source. Figure 1.4 presents the structure of a 9-level CHBMLI. By adding an individual H-bridge in cascade the output levels are increased by two. The maximum voltage levels that can be produced by the CHBMLI is given by:

$$V_{Level} = 2 \times N + 1 \quad (1.6)$$

Where N denotes the number of H-bridges used in the configuration.

CHB topologies are classified as Symmetric and Asymmetric Multilevel Inverter Vijeh *et al.* (2019). When the DC sources of a CHB inverter are similar in magnitude a Symmetric Multilevel Inverter (SMLI) is

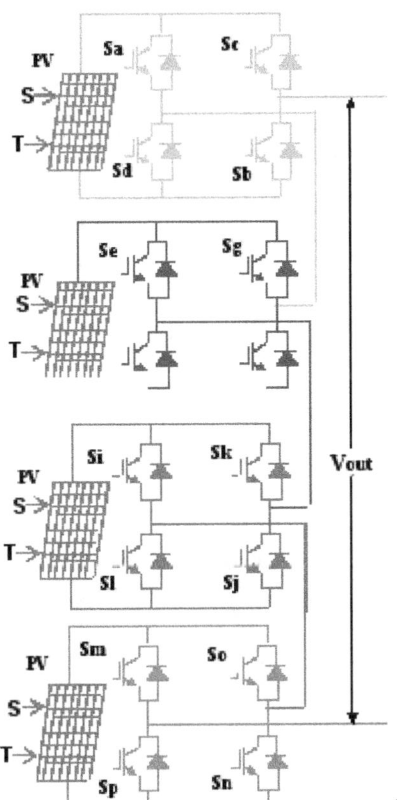

evolved. When the DC sources of a CHB inverter are distinct, then an Asymmetric Multilevel Inverter (ASMLI) is evolved.

Sh   Sf

S-Solar irradiation, T-Temperature

Figure 1.4 Circuit diagram of 9-level CHBMLI

The benefits of CHBMLI are:

- No additional components are needed, such as clamping diodes and flying capacitors.

- Similar to the two other multi-level inverters, the circuit architecture is very basic and lightweight for packaging.

- Can be operated using both symmetric and asymmetric DC sources.

- Great modularity and a wide range of redundant states.

The drawbacks of CHBMLI are:

- For real power transitions, it needs several independent DC sources.

- The lack of a mutual DC bus.

- CHB requires more devices to obtain back-to-back regenerative operation.

- More CHB cells must be used to improve the output voltage efficiency.

- The increase in H-bridges increases the quantity of transistors thereby increasing the losses of the system.

Diode Clamped or Neutral Point Clamped Multilevel Inverter

Nabae *et al.* (1981) introduced the first (NPC) Neutral Point Clamped MLI. A 9-level NPCMLI configuration is shown in Figure 1.5. The NPC is a fully developed configuration with a single DC source. This DC source is separated into several DC sources using capacitors. The NPC is

commonly used for all types of industries, varying from 2.3 kV to 4.16 kV, with few implementations up to 6 kV. In fact, NPC has found applications for

high-performance AC drives in oil, gas, and mining industries. NPC is also designated as Diode Clamped Multilevel inverter (DCMLI).

(Kumawat & Palmalia 2019) investigated the basic principles and functions of traditional MLIs. The DCMLI needs V_{Level} -1 DC-link capacitors for N phase voltage step.

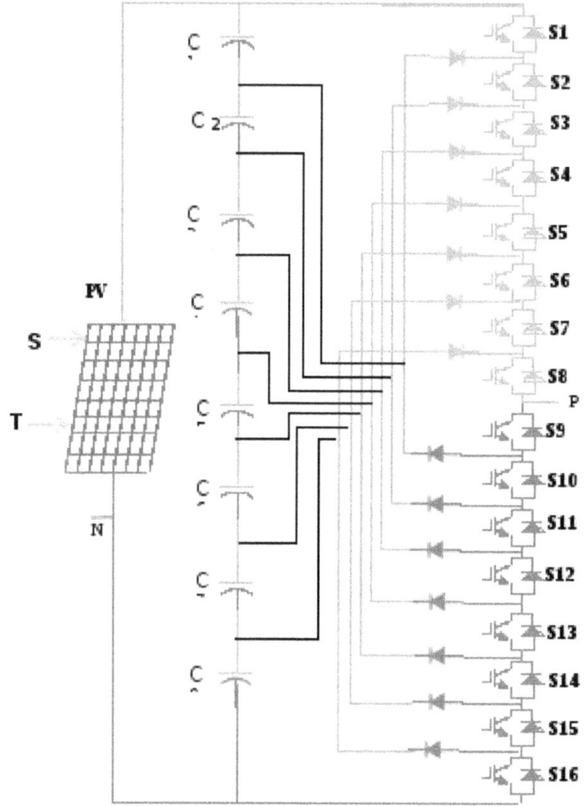

S-Solar irradiation, T-Temperature

Figure 1.5 Circuit diagram of 9-level DCMLI

The advantages of NPC will be as follows:

- Multiple DC-link capacitors are used with a single DC source.

- Easy regulation for regenerative action (back to back topology).

- Each switch blocks a voltage V_{dc}.
- A common three-phase DC bus that lessens the need of capacitors.

NPC has the following disadvantages:

- Individual regulation of real power flow in each leg of the phase is difficult.
- Devices with unequal rating leads to unequal distribution of load current in each device.
- High power diodes are required.

- An additional circuit is needed to balance capacitor voltage, thus increasing the inverter's complexity and cost.
- The number of clamping diodes required shall be increased if the level increases.
- The circuit design is more complex.

Flying Capacitor Multilevel Inverter (FCMLI)

Flying Capacitor Multilevel Inverter is a quite well-known topology of multilevel inverter. As the capacitors float in relation to the earth's potential the name flying capacitor is used. It is quite similar to a multilevel inverter clamped with a diode. It has one DC source and a group of switches, capacitors, and diodes. Figure 1.6 displays the circuit of a 9-level FCMLI. The circuit has been termed as flying capacitor MLI because multiple condensers clamp the system voltage to a single capacitor voltage. Each capacitor connected in series has the same voltage ratings.

The benefits of flying capacitors are:

- Extra driving energy during power out.

- Provides different switching paths.

- Controls real ad reactive power.

- The flying condensers have balanced DC voltages if the central DC bus capacitor voltage is balanced.
- It is possible to achieve independent capacitor voltage control.

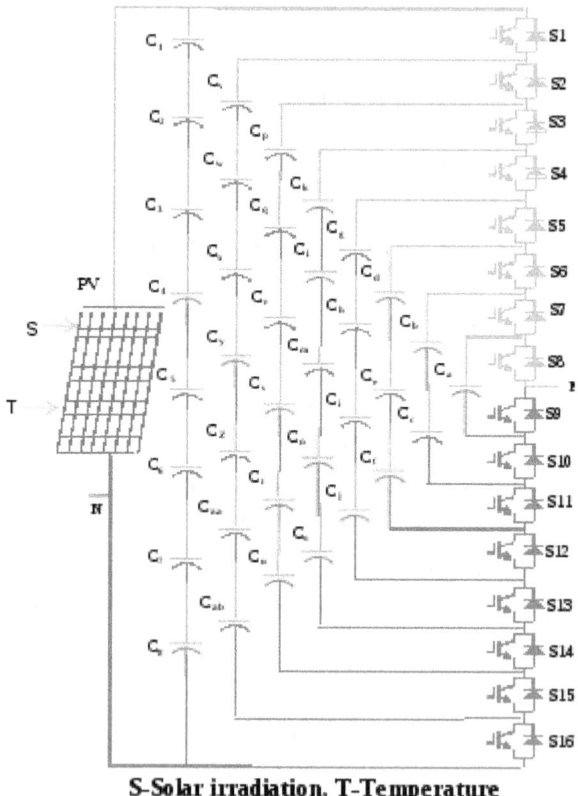

Figure 1.6 Circuit diagram of 9-level FCMLI

The drawbacks of flying capacitors are:

- Complex voltage control in real power conversion.

- The inverter is not effective due to numerous capacitors.

- It is tougher to pack and costly.

- High switching errors and complex inverter control.

- Not ideal for low frequency switching.

Drawbacks of Traditional MLI Topologies

Among the three traditional topologies, the NPCMLI is a well-known architecture due to its simplistic transformer power-circuit layout with a lower device count and it requires a lesser number of capacitors. The structure of the NPCMLI can be expanded to higher levels, but these are less desirable as the losses are greater and they are unevenly distributed. In addition, the series connections of clamping diodes increase the conduction losses and generates reverse recovery current during switching. This leads to higher switching losses in NPCMLI. In addition, the capacitor voltage balance is not achieved in high-level configurations with the passive front end when traditional modulation approaches are used. In this scenario, it is not possible to maintain the multilevel stepped waveform and to stabilize the capacitor

___voltages for some modulation indexes and therefore higher $\frac{dv}{dtls}$ are needed.

The CHB-MLI provides greater voltage performance with low voltage rated semiconductors for high-power applications.

CHBMLI needs a significant number of independent DC sources. These DC sources can be obtained from either RES or phase shifting isolation transformers. If phase shifting transformers are used, then the overall cost of the system increases since these transformers are costly and heavy in nature when compared with the conventional transformer. However, this topology has been useful for reducing the harmonic distortions in the current and increasing the input power factor of the converter. Even though the structure of FCMLI is modular, its industrial applications are very limited when compared with NPC and CHB. The major drawback regarding FCMLI is to balance the capacitor voltages. For balancing the capacitor voltages higher switching frequencies

more than 1200 Hz should be used. But the switching frequencies are

typically restricted to 500-700 Hz for high-voltage applications.

PROBLEM IDENTIFICATION

The investigation and advancement of multilevel inverters are essential to enhance the power quality in the power system. Due to the nonlinear characteristics of power converters the three-level inverters could not maintain the power quality within the IEEE grid code requirements. Therefore, on the output side of three-level inverters, LC filters are implemented to reduce the harmonics and improve the power quality. Although the stresses across the switches are usually not reduced, the stability and effectiveness of the system get deteriorated Sathik & Vijayakumar (2019).

The issues faced by the three-level inverters can be solved by implementing MMI. Because of the numerous advantages, MMIs are greatly preferred in medium voltage and high-power applications. The use of high switching frequency modulation methods creates voltage stress on the switches and affects the output voltage of the inverter. However, certain issues identified in traditional multilevel inverters and their modulation strategies are outlined as switch reduction, harmonic reduction and grid synchronization.

Switch Reduction

In traditional multilevel inverters, as the output voltage levels increase the switching components also increase. Since the number of switching components increases, the gate drivers associated with the number of switching components also increase. Therefore, triggering of switching components become complex and affects the reliability of the inverter.

Switching Method

Another important criterion for the enhancement of power quality at the output side is the modulation method in the multilevel inverter. The use of high switching frequency modulation methods produces losses in the inverter thereby decreasing its efficiency. By implementing high switching frequency modulation techniques, the harmonic distortions present in the output voltage can be reduced, but high switching losses occur during switching transitions.

Grid Synchronization

Designing an adaptive controller is one of the issues which have to be considered in a grid-connected system (GCS). Due to the lack of efficient controller voltage swells, voltage sags, power fluctuation occurs in the PV system, and lack of reactive power compensation arises in a GCS. These problems lead to poor voltage regulation and instability. Therefore, there is a need for an adaptive controller to pump harmonic less power to the grid and compensate reactive power.

These issues related to the three-level inverter and traditional MLI can be overcome by implementing the proposed MMI.

OBJECTIVES OF THE THESIS

From the background, it can be observed that multilevel inverter topology gets high importance in the area of power electronics systems. Hence, the objectives of the proposed work are as follows:

- To implement the P&O MPPT algorithm for the PV system in both shadowed and non-shadowed conditions, whose variables comply to the hardware specifications

- To design a symmetric 9-level and asymmetric 27-level ,31- level inverter structure by reducing the switching components

- To design and develop the SHE PWM method for a PV fed Symmetric Modular Multilevel inverter (SMMI). Optimization methods such as Particle Swarm Optimization (PSO), Genetic Algorithm (GA), and Bee Optimization (BO) are used to compute the complex equations of SHE. Moreover the $3_{rd}, 5_{th}$ and 7_{th} order harmonics are selected to be eliminated.

- To design and implement an adaptive closed-loop control for the suggested solar fed modular multilevel inverter since conventional controllers have slower response to the transients. The methods include PSO-PI, FFA-PI, and HHO PI-based controllers

The proposed structure and the modulation technique is simulated with MATLAB/Simulink (R2017a) software. The hardware implementation is carried out with Field Programmable Gate Array (FPGA) based processor to generate the switching sequences based on the proposed methodologies.

INTRODUCTION TO MODULAR MULTILEVEL INVERTER

MLI is focused on the creation of novel structures, modulation techniques, and control schemes. The introduction of MLI's indicated maximum interest from both industry and academia. In addition to traditional topologies, researchers tend to create modern topologies for an application- oriented approach. In fact, the implementation of MMI (Modular Multilevel Inverter) is an significant contribution to the advancement of the MLI topologies. It has become an effective topology in medium or high voltage applications.

A configuration with a reduced part count is implemented to overcome the issues faced by conventional topologies. While several structures based on the reduced part count have been implemented, modular configurations still remain to be developed. Generally, design of novel architecture incorporates three strategies, such as structural design improvement, the introduction of uneven or uneven DC sources rather than even or equivalent DC sources and the integration of design improvement with asymmetric DC sources.

The most common terms in MLI configuration is the symmetrical and asymmetrical. Both the configurations have their own advantages and drawbacks. In contrast to the SMLI, the ASMLI generates more output levels with the same number of components. The ASMLI not only generate higher voltage rates but decreases the difficulty in switching states. Binary MLI and trinary MLI are the most common configurations of ACMLI. Asymmetric reduced MLI switches are often split into two main sections, such as with H- Bridge and without H-Bridge.

MLI is compensated to emphasize the creation of new configurations, modulation methods, and control schemes in the understanding of their great advantages. But traditional MLI requires several switching devices to increase the voltage levels. As the switching device increases the demand for driver circuits, passive heatsinks, and safety circuits also increase. As a result, the size of the traditional configuration becomes bulky, expensive, and difficult to control.

DESIGN OF SOLAR PV FED MODULAR MULTILEVEL INVERTER

Design of PV System

PV design is divided into two major models; the first is a single diode, which is the easiest type, and the second is a double diode, which is a complex form of representation than a single diode. Though single diode is more common for PV modeling, it has several drawbacks such as

- PV performance exhibits high deficits while operating under variable temperatures.

- Recombination loss is neglected in the PV cell depletion region.

- At low irradiance levels the accuracy of the system is deteriorated especially at open circuit (V_{oc}).

The two-diode model is used in the proposed work since it has increased PV device performance and improved curve fitting. The single- diode configuration is not considered because of its drawbacks.

Solar PV cell architecture involves the following equations:

$$I = I_{solar} - \mu_1 - \mu_2 - (V_s + 1^* R_s)/R_p \quad (1.7)$$

Where

$$\mu_1 = I_{ds}^* (e^{E^*th} - 1) \quad (1.8)$$

$$\mu_2 = I_{ds2}^* (e^{E2^*th} - 1) \quad (1.9)$$

Where I_{ds} and I_{ds2} denotes the diode saturation currents, V_{th} specifies the thermal voltage of the solar cell, the emission coefficients $£_1$ & $£_2$ are related to the diodes D_1 and D_2 and I_{solar} is the current produced by the PV

cell.

$$I_{solar} = I_{cr}^* (I_{so}/I_{cro}) \quad (1.10)$$

Where, the solar current is indicated as I_{so}, denotes and the irradiance current is indicated as I_{cro}. PV systems are equiped with DC-DC

converter to regulate the voltage obtained from PV panels as discussed by Kishore & Bhima Singh (2020).

Design of Modular Multilevel Inverter

The suggested topology of MLI consists of a limited number of unidirectional switches (6), bidirectional switches (4), and few DC sources (5). These devices are smartly designed to provide the desired level of the output voltage. The purpose of the bidirectional switch is to obstruct the positive or negative voltages and allow currents to pass in either direction. Since the common emitter structure has less conductivity loss and needs only one gate driver circuit, it is selected for implementing in the suggested topology. Depending on the ratio of input DC sources the suggested inverter can be operated in symmetric and asymmetric modes.

Figure 1.7 shows the proposed MMI fed with PV sources. Since five separate DC sources are used, the proposed inverter is ideal for solar PV applications. The fifth DC source is implemented to produce the negative voltage levels. It is obtained by summing up the other four individual DC sources. This topology does not require the use of any kind of auxiliary circuit for the generation of negative levels. The use of H-bridge for producing the negative level is reduced. When the DC sources are identical in magnitude, symmetric multilevel inverter is obtained. The ratio of DC sources for

symmetric multilevel inverter is 1: 1: 1: 1 and the fifth DC source (V_{peak}) is

obtained by adding the four individual DC sources. Therefore, for the symmetric 9-level inverter the fifth DC source is 4 V_{DC}. The symmetric multilevel inverter generates 9-level output voltage with six unidirectional switches and four bidirectional switches. The MMI can achieve additional output voltage levels (11,13,15,17,19,21,23,25,27,29,31) using a variety of asymmetrical DC sources.

The number of switches is equivalent to the number of driver circuits in the topology even though it utilizes unidirectional and bidirectional switches. The proposed configuration operates at both symmetric and asymmetric conditions without changing the structure of MLI. The MMI consists of four solar PV modules connected in series,which are labeled as PV_1, PV_2, PV_3, and PV_4. The power obtained from the PV modules is being transmitted to the inverter circuit consisting of unidirectional and bidirectional devices.The switches in the MMI are designated as S_1, S_2, S_3, S_4, B_1, B_2, B_3, B_4, $P_1$and P_2. Table 1.1 describes the equations of the proposed methodology based on the levels,switches,diodes,drivercircuits and DC sources.

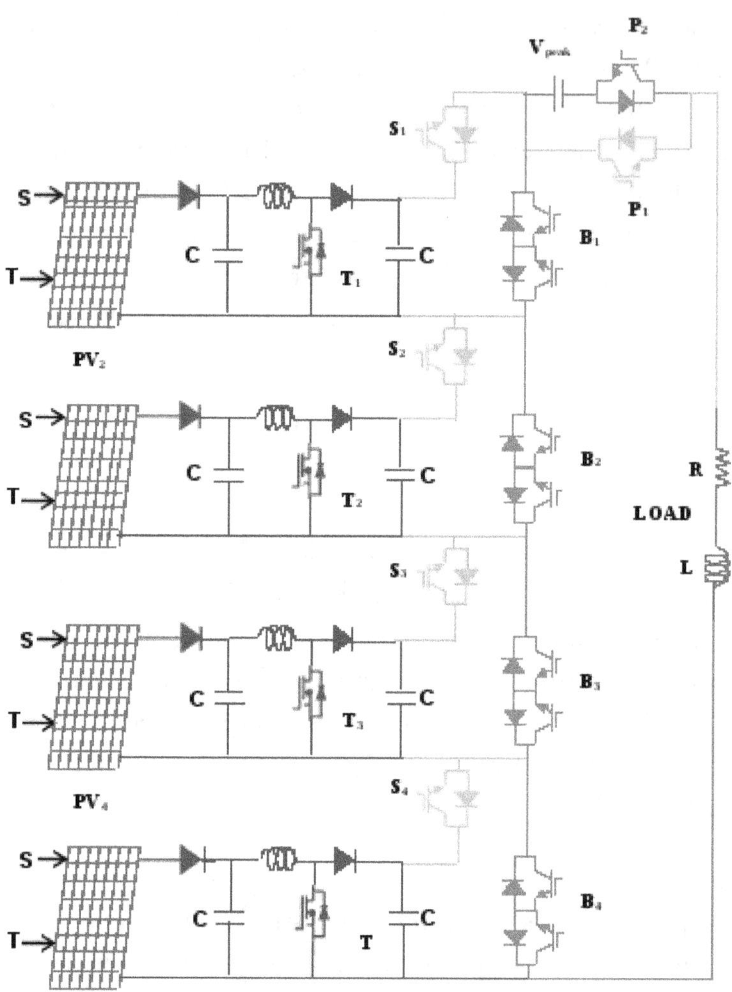

Figure 1.7 Proposed Solar PV Fed Modular Multilevel Inverter

Table 1.1 Equations of the Proposed Symmetric And Asymmetric Modular Multilevel Inverter

Parameters	Symmetric	Asymmetric
Number of levels	N_L(9-level)	N_L(27-level)
Number of unidirectional switches	$(N_L - 3)$	$\dfrac{N_L + 3}{5}$
Number of bidirectional switches	$(N_L - 4)$	$\dfrac{N_L + 3}{6}$
Number of diodes	$(N_L + 5)$	$\dfrac{N_L + 1}{2}$
Number of driver circuit	$(N_L + 5)$	$\dfrac{N_L - 7}{2}$
Number of DC sources	$(N_L - 4)$	$\dfrac{N_L - 7}{4}$

The DC-DC converter controls the voltage supplied to MMI which has a solar photovoltaic module with a node voltage equal to reliable power. The power produced by the given Photovoltaic module completely depends on solar irradiance $(1000 \text{W}/\text{m}_2)$ and temperature (25°C). In fact, the solar PV module's nodal voltage should be equivalent to the resultant.

There is no repetitive switching sequence for producing each output voltage step. Since the proposed structure doesn't require any extra circuit for polarity generation part the voltage stress on the H-bridge switches are absent. Moreover, the voltage stress on the switches of this topology is equally distributed. Consequently, the proposed topology has the potential to operate in medium voltage applications by selecting the Insulated Gate Bipolar Transistor (IGBT) switches.

Figure 1.8 illustrates the output waveform of a 9-level multilevel inverter. Although the number of levels increase, the output will not be an exact sine wave. So, the concept of "Total Harmonic Distortion (THD)" is used to characterize the nature of the output voltage of a non-sinusoidal wave. Equations 1.11 and 1.12 are used to calculate the THD of the current and voltage waveforms.

$$\text{THD}(V) = \frac{\sqrt{\sum_{q=2}^{\infty}\left(V_{q,\text{rms}}\right)^2}}{V_{f,\text{rms}}} = \frac{\sqrt{(V_{o,\text{rms}})^2-(V_{f,\text{rms}})^2}}{V_{f,\text{rms}}} \tag{1.11}$$

$$\text{THD}(I) = \frac{\sqrt{\sum_{q=2}^{\infty}(I_{q,\text{rms}})^2}}{I_{f,\text{rms}}} = \frac{\sqrt{(I_{o,\text{rms}})^2-(I_{f,\text{rms}})^2}}{I_{f,\text{rms}}} \tag{1.12}$$

Where $V_{q,rms}$, $I_{q,rms}$ are the RMS voltage obtained at q_{th} harmonic,

$V_{f,rms}$, $I_{f,rms}$ are the RMS voltage/current of the fundamental component and

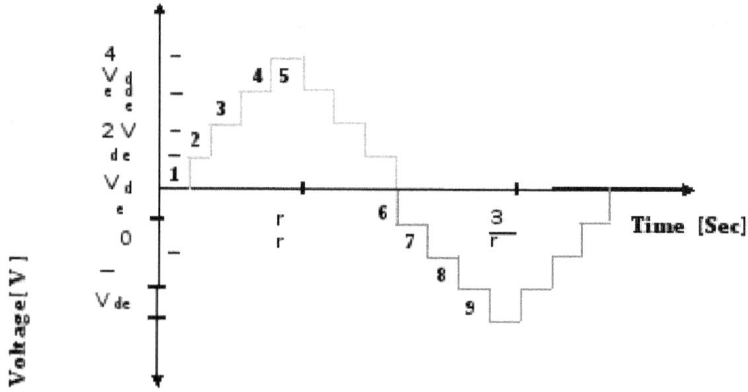

$V_{o,rms}$, $I_{o,rms}$ are the RMS voltage/current of the output value. Based on the load the current THD varies.

Figure 1.8 Nine-level output voltage waveform

The reduced switch topology is compared with traditional topology to generate better results, making it more attractive for solar PV applications. Since DC-link capacitors are not essential for the suggested configuration, the problem of voltage balancing is absent. The efficiency of the MMI is also determined by the on state switches during conduction. To generate any voltage level from the proposed structure, only 5 switches conduct at each instant. Therefore, out of fourteen switches only five switches are in conduction all the time to produce the respective voltage levels.

CONTROL STRATEGIES FOR MODULAR MULTILEVEL INVERTER

The output voltage of MLI contains unwanted harmonics. Harmonics are characterized as multiple integer frequencies of the basic frequency 50Hz presented either in voltage or in current waveforms. The harmonics are often produced on the source side or on the load side. Nonlinear loads such as arc welding, power converters, personal computers, and lighting cause harmonics. The non-sinusoidal current creates more harmonics, and these harmonics cause overheating in the winding of the transformer and machines. The output harmonic content in an MLI mostly depends upon the switching angles of the power converter.

To suppress the harmonics many modulation schemes have been introduced by the researchers in the past few years. The implementation of a high switching frequency modulation technique for MLI will damage the switches by producing a thermal loss. Therefore, low switching frequency modulation techniques are mostly implemented for harmonic elimination in MLI.

The Pulse Width Modulation (PWM) technique is one of the well- known modulation techniques, which generates low THD in the output voltage, but the switching losses are larger due to the higher switching frequencies. Different PWM techniques utilized within the multilevel inverters are the sinusoidal PWM technique Hossam-Eldin *et al.* (2020), Space Vector PWM (SVPWM) Wu *et al.* (2020), and Selective Harmonic Elimination PWM (SHE-PWM) Routray *et al.* (2020).

The sinusoidal PWM technique generates a common mode voltage in motor drive applications due to the higher switching frequency, and this will damage the motor bearing and motor winding insulations because of high dielectric stress, circulating current in power circuits, voltage or current transients, and corona discharge (Kant & Singh 2020).

SVPWM technique is the advanced modulation technique, which functions in low switching frequency in terms of hundreds of hertz. The switching pattern and the setup of the hardware are more complicated because of more computation for vector calculation and it needs an enhanced digital controller like a digital signal processor or Field Programming Gate Array (FPGA).

In low switching frequency modulation technique, the selective harmonic eliminating pulse width modulation (SHEPWM) suits best for removing the low order harmonics from the inverter output voltage. The issues related with SHE-PWM method (Kulothungan *et al.* 2020) is to solve the nonlinear transcendental equations, and this can be fixed by numerous computational methods such as a genetic algorithm (Wang & Sobey 2020), bee colony optimization Falehi (2020), and cuckoo search (Singla & Sharma 2020). This encouraged the focus on developing a novel MLI with reduced power switches and to bring in a fundamental switching technique for high- power applications. In this work, the firing angles for the switches of MMI are obtained by solving the complex transcendental equations. These nonlinear trigonometric equations which are complex in nature can be solved using evolutionary algorithms (Panda *et al.* 2020).

Calculation of Switching Angles using Optimization Techniques

The optimization techniques are evolved by considering the biological behaviour of birds and animals. These algorithms have several benefits such as faster response, adaptiveness in obtaining a perfect solution, and a higher chance of convergence. The proposed work uses three biological algorithms namely Particle Swarm Optimization (PSO), Genetic Algorithm (GA), and Bee Colony Optimization (BCO). These methods use an objective function which contains the complex trigonometrical equations of the fundamental and low-order harmonics. In addition, these strategies reduce the objective function to achieve the best possible firing angles to remove the undesired harmonics.

In the proposed work, a 9-level is obtained by using 4 symmetric DC sources. The fifth DC source is responsible for producing the negative levels. In SHE PWM the triggering angles play an important role for generation of output voltage levels. The triggering angles required to obtain the desired output levels is given in Equation 1.13.

$$= \frac{N_{Level} - 1}{2} \tag{1.13}$$

Where S_{ang} denotes the number of switching angles, N_{Level} denotes the number of levels.

The harmonic order (H_{OD}) is expressed as

$$H_{OD} = (S_{ang} - 1) \quad (1.14)$$

Therefore, four triggering angles (θ_1, θ_2, θ_3 and θ_4) are essential to

generate a 9-level output for the suggested MMI. Based on the Fourier series expansion, four nonlinear equations are obtained. The primary equation is equated to the fundamental component and the other three equations are equated to zero. These equations are computed using the optimization techniques such as GA, PSO, BCO to obtain the switching angles and are stored in the lookup table for reducing the $3_{rd}, 5_{th}$ and 7_{th} order harmonics.

Bee colony optimization

In this method, the finest solution in solving a problem is the food system. The food system is a N dimensional function in which N provides the number of variables for optimization. The fitness value can be measured by the quality of honey in the food system. Bee algorithm implements three steps such as the employed bee phase, onlooker bee phase, and scout bee phase to produce the triggering angles for the proposed inverter.

Genetic algorithm

A genetic algorithm is a heuristic method influenced by the theory of natural evolution of Charles Darwin. In this method, with the appropriate choice of population size, selection mechanism, cross over and mutation rate, stopping criteria, the optimal values of switching angles are obtained. These optimal switching angles are directly fed to the proposed inverter using a programmable pulse generator.

Particle swarm optimization

In this method, for the given objective function and constraints, the position and velocity are initialized. After the updation of velocity and calculating the new position, the optimal value (G_{best}) is obtained. On fixing the stopping criteria, the optimal values of switching angles in the form of G_{best} is achieved for the proposed inverter.

CONTROL STRATEGIES FOR GRID CONNECTED MODULAR MULTILEVEL INVERTER

The proposed grid-connected photovoltaic system contains photovoltaic array, DC-DC boost converter, MMI, and the grid. The MMI is inked with the grid through the Point of Common Coupling (PCC). The MMI plays a major role in interfacing the PV system with the grid. The Perturb and Observe (P&O) MPPT method is used to trap a maximum amount of energy from the PV panels. For integrating PV system to grid a two-stage conversion process is needed. In the first stage, DC-DC converters are implemented for boosting the low voltage obtained from PV panels to a higher voltage and in the second stage, the boosted DC power is converted to AC power using MMI. Usually, PWM techniques are implemented to control the performance of the inverter.

In the closed-loop control systems, the commonly used controller is the Proportional and Integral (PI) controller. The ultimate goal of the proposed controller is to pump clear sinusoidal current to the grid even when there is distortion in the grid voltage or operation of a nonlinear load or any voltage imbalance. Since a 31-level AMMI is used for the grid connected system the input voltages fed to the AMMI should be in the ratio of (1:2:4:8).

The proposed closed loop control system consist of six control loops, of which four independent voltage control loops are used to balance the individual DC link voltages, an overall voltage control loop is used to balance the overall DC link voltage and a current control loop is used to control the inverter current which is then injected to the grid. The gains of PI controller present in the total voltage controller block and the current controller block in the control system are tuned using optimization methods such as PSO,FFA and HHO. In the overall voltage control loop the reference value is compared with the measured value using a comparator and the error is fed to the PI voltage type controller.

The output of the overall voltage controller is the inverter current with maximum amplitude. The amplitude of the inverter current can be adjusted by adjusting the gains of the PI voltage type controller. The inverter current obtained from the overall voltage controller is multiplied by the sine wave signal obtained from the phased lock loop (PLL) to obtain the reference grid current. To maintain the inverter current in phase with the grid voltage and to produce the sine wave signal from grid voltage for grid synchronization a PLL is used.

The controller is built based on the principle of power balance between the input and output of the inverter. The instantaneous grid current is equated with the reference grid current to feed ripple free sinusoidal current into the grid. The resulting error is given as the input for the PI current type

controller. The PI controller gains (K_p, K_i) are calibrated to maximize the

instant grid current equivalent to the reference grid current. In the proposed work, the performance of the grid-connected MMI system is examined using the PSO-PI controller, FFA-PI controller, and HHO-PI controller.

ORGANIZATION OF THE THESIS

This thesis is organized into six chapters reporting the complete research work where it deliberates and scrutinizes the outcomes.

Chapter 1 indicates the introduction to alternate energy sources, explains the issues faced by conventional configurations, problem statement and the aims of the study. The organization of thesis is also mentioned in this chapter.

Chapter 2 analyzes the literature on the key topics of this research such as traditional topologies of MLI, reduced MLI topologies, various modulation techniques, and optimization techniques employed.

Chapter 3 projects the design of the suggested symmetric and asymmetric modular multilevel inverter topology. The proposed structure has a high level to switch ratio and can produce 11 different levels of output voltage using a variety of DC sources. Comparison of symmetric 9-level MMI and Asymmetric 27 and 31-level is carried out in this chapter.

Chapter 4 describes the harmonic elimination of MMI using the selective harmonic mitigation method. The nonlinear transcendental equations of the SHE PWM technique are very complex to solve. Therefore, optimization methods such as BCO (Bee Colony Optimization), GA (Genetic Algorithm), and PSO (Particle Swarm Optimization) are implemented to optimize the switching angles and enhance the power quality of the system with reduced magnitude of selected harmonic orders. The proposed method is tested in a 1kWp solar PV plant with FPGA based processor.

Chapter 5 presents the grid connection of PV fed MMI. An optimization-based PI controller is employed for imparting ripple free sinusoidal current to the grid. The performance of the MMI using various controllers (PSO-PI controller, FFA-PI controller, and HHO-PI controller) are observed in steady-state and transient conditions.

Chapter 6 summarizes the work presented in the thesis and highlights the future work along with conclusions.

CHAPTER 2 LITERATURE REVIEW

INTRODUCTION

Multilevel power conversion technology originally arises from the need to reduce the total harmonic distortion in the inverter's output. Later, it is introduced functionally to satisfy the need for high-voltage DC bus-driven inverters. The switching devices of conventional inverters experience high voltage stress or lack sufficient voltage rating to produce high voltages. The stress on each switching component can be reduced proportionally by using multilevel structures. As a result, a high voltage DC bus can be handled easily without the use of heavy step-up or step-down transformers. Hence, MLI can be used without the bulky transformers in grid-connected renewable energy applications.

This chapter discusses the review of Modular Multilevel Inverter (MMI), modulation methods and about Grid connected systems. The literature review of power quality improvement techniques with a brief description of the modulation strategies, optimization techniques of an inverter is shown in Figure 2.1. The need to produce electricity by RES, such as solar and wind energy, is primarily to reduce pollution with the lowest maintenance costs and effectively integrate the MLI.

The quality of the output will be considered as one of the key challenges from electrical suppliers to end consumers in order to improve electrical efficiency. As a result, many researchers have focused their attention on MLI to increase the quality of electrical power by minimizing harmonics.

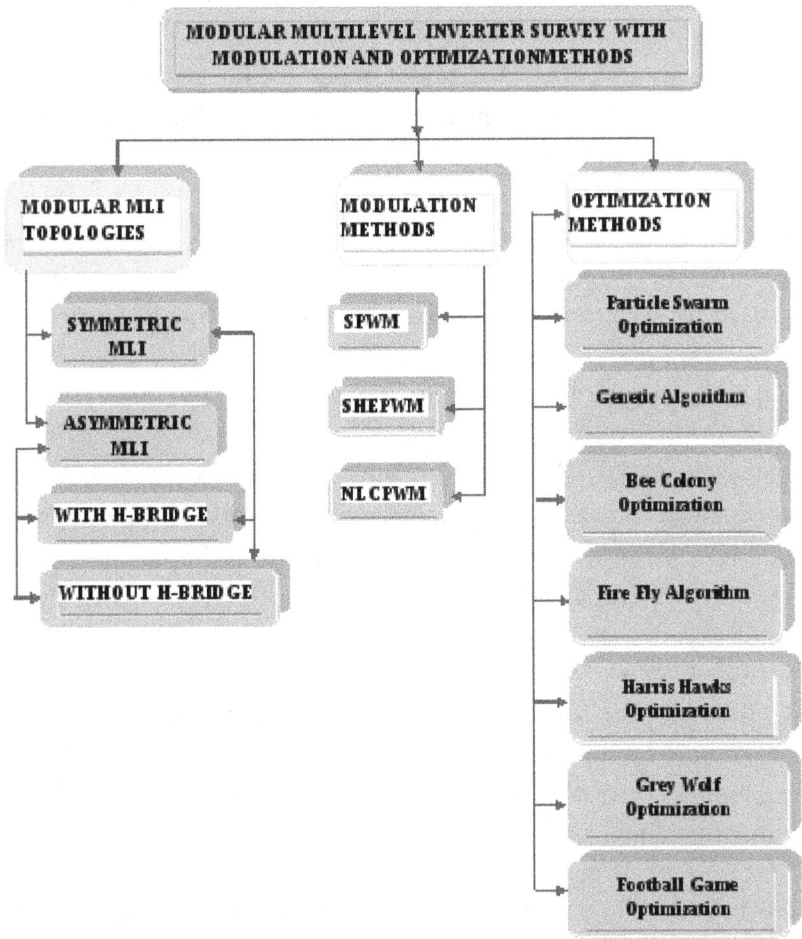

Figure 2.1 Survey on MMI with Modulation Methods

LITERATURE REVIEW OF SOLAR PHOTOVOLTAIC SYSTEM

Singh *et al.* (2020) proposed a multipurpose distributed sparse (DS) control approach for a photovoltaic system. The proposed system uses the P&O method for harvesting maximum power from PV cells. In the absence of a PV system the suggested system operates as a DSTATCOM.

Mishra & Singh (2020a) proposed a standalone water pumping system based on solar energy. Switched Reluctance Motor (SRM) is used for water pumping in the proposed system. The proposed water pumping system uses the P&O method for harvesting maximum power from PV cells.

Jhang *et al.* (2020) designed a boost DC-DC converter and is constructed using Complementary metal–oxide–semiconductor (CMOS).A low leakage logic gate circuit is implemented for lowering the initial voltage.

Dehghani *et al.* (2020) pointed out the advantages of the P&O method. The performance of solar PV system in a grid connected network is tested using the P&O based controller and the Fuzzy controller. The results show a moderate performance of the fuzzy controller when compared with the P&O method.

Mishra & Singh (2020b) proposed a solar powered grid integrated water pumping system .The authors used the single stage conversion for the solar PV system. The motor used for water pumping is a switched reluctance motor. For harvesting maximum power from the PV system a P&O method is used. Since the PV systems are non linear in nature and its performance is mostly depending on the climatic conditions the P&O method suits best for tracking maximum power under any conditions.

Bhattacharyya *et al.*(2020) proposed a fast tracking method for P&O, Incremental Conductance (InC) methods to obtain constant output from the PV system. Implementing a flexible step size for both the methods tracks the maximum power point very fast and delivers a stable output. Compared to the Zero Oscillation based P&O algorithm (ZA) and Device Duty Ratio based P&O algorithm (DDR) the flexible step size based P&O method performs best.

Erauskin *et al.* (2020) presented a multi-variable P&O method for distributed MPPT system. The main advantage of this type of P&O method is that it can effectively function by lowering the voltage stress on the DC-DC converter and it perform well during the partial shaded conditions.

Ram *et al.* (2020) proposed an enhanced leader particle swarm optimization (ELPSO) for the P&O method to tract the maximum power under any adverse conditions. The global maximum power point (GMPPT) is identified by using the enhanced leader particle swarm optimization method. Using the enhanced P&O method better and accurate results are obtained.

Mishra *et al.* (2020) proposed a drift free P&O method for the stand alone wind energy conversion system. The conventional P&O method provides oscillations and cannot tract the MPP at varying wind speed. Using the speed sensor the wind data is being obtained and given to the proposed P&O method to extract accurate power.

Hosseini *et al.* (2020) conducted experiments on PV panels covered with snow. The performance of the PV panel under non-uniform snow accretion is examined. In addition, the use of PV panel bypass diodes is examined. Finally, the efficiency of various PV panel designs is tested and further studied under cold weather conditions. The author's work can be used as a tool for designing PV systems in cold climatic regions.

Schmitz *et al.* (2020) presented a survey on step up DC-DC converters. Various high gain DC-DC converters were reviewed on the basis of their device rating, switching frequency, voltage stress, inductor size and control of output voltage. All the configurations were experimentally tested and their performance were analysed.

Aroudi *et al.* (2020) conducted investigations on boost converter using digital controller. While operating at constant power load, the DC-DC boost converter produces inrush current during its start up. This inrush current is because of the computational delay of the digital controller. A slope limiter is used to minimize the inrush current of the boost converter.

LITERATURE REVIEW OF CONVENTIONAL

MULTILEVEL INVERTER TOPOLOGIES

Diode Clamped MLI

DCMLI includes a number of switches which are aligned in a sequence with the clamping diodes. DCMLI has the disadvantages of

1. external internal clamping interface and (2) several diode blocking voltages. The limitations of DCMLI are the voltage balancing issue across the DC bus capacitors, higher-clamping diodes improve the system's strength, as it becomes impossible to stabilize the capacitor when the number becomes greater than three.

Hossam-Eldin *et al.* (2020) implemented a DCMLI in Wind Energy Conversion System. The Double Fed Induction Generator (DFIG) is being operated by the DCMLI. To produce ripple free power to the DFIG, Sinusoidal Pulse Width Modulation (SPWM) and third harmonic injection pulse width modulation (THIPWM) techniques are implemented for the DCMLI.

Flying Capacitor MLI

The comparison between DCMLI and FCMLI topologies differs in the usage of the flying capacitor to replace the clamping diode. The primary modification between FCMLI and DCMLI. The inverter design is identical to the DCMLI under which the capacitors are used rather than clamping diodes. The FCMLI requires a sequence of switching cells . The key challenges for FC inverter are: (1) need of greater amount of capacitors, (2) the usage of large quantities, rendering the system more expensive, and (3) the correct option of switching modes avoids the question of voltage balancing as the amount of levels is greater than five.

Azurza *et al.* (2019) introduced a modified seven-level FCMLI, setting a new standard for ultra-highly efficient and powerful converters. Substantial volume savings can be accomplished if the number of FC cells is halved, especially in the case of low frequencies, as the power needed to protect the flying capacitor's ripple is inversely proportional to the frequency of switching, which is constrained by ultra-efficient converters. However, because of the high device performance and the losses between many switches and/or control supplies, the design effort is reduced.

Cascaded H-Bridge MLI

CHBMLI is associated with various sources of voltage to obtain a greater amount of output voltage levels. It enables both addition and subtraction of the switching state combinations between the nearby DC inputs to produce the specified output. The inverter will produce variable output voltage levels using variable DC sources. Increasing H-bridge in the structure increases the output voltage levels. The restriction of this configuration is that for every H-bridge connection, a separate DC source is required.

Jahan *et al.* (2019) suggested a CHB topology, which uses a single source and switched capacitor, instead of multiple DC sources. For capacitor charging, a single DC source is needed. The capacitor charging cycle in a switched capacitor cell usually involves some current spikes that cause great harm to the charging device and the capacitor. An auxiliary circuit is implemented to reduce the spikes during the charging of the capacitor. The auxiliary circuit,16 switches,4 capacitors, and a single DC source produces nine levels of the output voltage.DSP controller with LS-SPWM is used to produce the triggering pulses to the proposed structure.

LITERATURE REVIEW ON MODULAR MULTILEVEL INVERTER

Modular multilevel inverters were designed since it has advantages like reduction in the device count, reduced blocking voltage across the switches, capable of producing symmetric and asymmetric configurations, even power distribution, and modularity Shuvo *et al.* (2019). In traditional MLI the switches are being increased to produce high voltages across the output terminals with less harmonic content. Each switch of a traditional MLI needs a gate driver circuit, safety circuit, and heat sink. This makes the traditional MLI costly and bulky. As a result, attempts were made in recent years to reduce the switching devices, and several topologies were listed in the literature. Each structure has its own benefits and drawbacks.

Omer *et al.* (2020) have examined the existing reduced switch MLI topologies. Moreover, the review provides a list of appropriate MLI topology references and control strategies. Although the improvement of topologies has been accompanied by a development in modulation schemes

A detailed overview of several modular multilevel inverter technologies, ideally adapted for different uses like motor drives, FACTs, and renewable power systems, is discussed in this section. These configurations may be found in both grid-tied and standalone applications. In general, MMI topologies are classified into two classes, i.e. Symmetrical MMI (SMMI) and Asymmetrical MMI (AMMI).

Symmetrical Modular Multilevel Inverter with H-Bridge

The word symmetric means that the magnitude of all the DC inputs used in a circuit are similar.

Dewangan *et al.* (2019) have provided two configurations of reduced MLI structures with fault tolerance features. One control switch is suggested to be attached properly to a 5-level (Cross Connected Source) CCS- MLI and T-Type MLI topologies to allow them to accept defects. Both updated structures have been called fault tolerant CCS-MLI and fault tolerant T-type MLI, respectively. It is observed that under any specific switch open circuit fault the MLIs operate satisfactorily.

Gandomi *et al.* (2019) proposed three separate methods for calculating the intensity of DC voltage sources and transformer turn ratios to produce even and odd voltages at the output. The basic inverter design is also established further and two new modular MLI systems were obtained. The systems built to raise the input voltage without an external boost converter. The additional benefits of the suggested topologies involve reduced power switches, peak inverse voltage (PIV) and DC voltage sources

Siddique *et al.* (2019a) proposed the concept of MLI topology depending on switched-capacitor. A novel multilevel boost structure in inverter topology is implemented. The topology proposed uses 11 unilateral switches for producing a 9-level output, with a single switched capacitor device. The main advantages of this topology are self-voltage balancing of the capacitor without any auxiliary system and reduced stress. For switch control, pulse width (PD-PWM) technique of phase disposition and nearest level control PWM (NLC-PWM) were used.

Sandeep *et al.* (2019) proposed a symmetric 5-level boost inverter with reduced switches. It requires 9 switches to generate a 5-level of output with one DC source and one capacitor. This topology is highly suitable for solar photovoltaic applications.

Salehahari *et al.* (2019) suggest a 9-level single-phase transformer configuration. The proposed inverter produces nine-level of output using only one DC source and eight switching elements. The voltage stress is equally distributed on the switches. The suggested arrangement will enhance the voltage without an auxiliary boost device. Another aspect of the suggested system is its extensibility to produce higher levels of output. In fact, in contrast with the analogous systems, the new system has low cost and length.

Lee *et al.* (2018) proposed a new hybrid cascaded MLI with a 4-level submodule, which substantially decreases the switch count to reduce cost. The SHE-PWM solution has been used for MLI control. This suggested structure can be used in photovoltaic and fuel cells for renewable energy generation.

He & Cheng (2018) presented an innovative MLI based on switched-capacitor. SPWM method is implemented for the suggested structure. The PWM method is used to improve efficiency and power density in this modular structure.

Panda & Panda (2018) suggested a novel reduced part count MLI. This topology needs 50% lesser switches than the CMLI. The harmonic spectrum can be enhanced by utilizing filters, but the effects of the dominant harmonics are still very important. By evaluating optimum triggering angles using the SHE method, the author has tried to remove the harmonic distortions. The author created an improved Particle Swarm Optimization (PSO) strategy to obtain the triggering angles. This configuration can be applied to high and medium power applications.

Lee (2018a) suggested a new hybrid CMLI with a single DC source and two capacitors to produce a 7-level of output voltage. Two bidirectional switches are used to integrate the capacitor unit. The LSPWM solution has been used for MLI control. This suggested structure can be used in photovoltaic and fuel cells for renewable energy generation.

Symmetrical Modular Multilevel Inverter without H-Bridge

Siwakoti *et al.* (2020) proposed a novel seven-level MLI with 8 switches and two FC. The number of switching elements is reduced in this special arrangement. A simple modulation method is implemented for the suggested structure to reduce the cost and complexity of the control system.

Liu *et al.* (2020) developed a 7-level MLI with nine switching devices. The author built the structure from a switched capacitor and a hybrid MLI structure. The proposed structure doesn't use H-bridge for producing negative voltage levels. The proposed structure is efficient and costs less.

Mahato *et al.* (2020) introduced a 9-level SMLI with 10 power switches and 5 DC sources. The strategy employed is modular and does not include H-bridge for generation of negative levels. SHEPWM technique is used to provide pulses to the proposed MLI.

Roy & Sadhu (2020) suggested the MLI using switched capacitor (SC), which generates 13-level using 14 power switches and 6 DC sources.

No additional circuit is needed to produce negative voltage levels in the designed structure. The proposed topology produces the lowest cost function when compared with other capacitor switched MLIs.

Lee & Lee (2020) designed a 7-level inverter with three-H bridges, two bidirectional switches, two capacitors, and one DC source. The proposed structure doesn't require any auxiliary circuit to produce negative levels. The voltage stress is equally distributed among the switches.SPWM method is implemented for triggering the power switches of the suggested topology.

Bhatnagar *et al.* (2019a) proposed a 9-level switched capacitor MLI with a single DC source including 2 capacitors. The 9-level MLI introduced guarantees a reduced device count. Furthermore, two of the eleven switches function at the standard switching frequency, thereby decreasing the switching losses. The topology suggested is triggered by a multicarrier PWM method.

Bhatnagar *et al.* (2019b) proposed a modified version of single stage switched capacitor, which contributes to a substantial decrement in the device count, with a corresponding decrease in overall peak inverse voltage. In the symmetrical configuration, the proposed topology will produce 9 voltage levels and in asymmetrical configuration 11 voltage levels is obtained using eight semiconductor switches, two DC sources, and two capacitors . Moreover, the switching losses are reduced by using standard switching frequency.

Lee *et al.* (2018b) proposed a 7-level and a 13-level MLI, which substantially decreases the device count to reduce costs and THD. The SHE- GA PWM solution has been used for MLI control. This suggested structure can be used in photovoltaic and fuel cells for renewable energy generation.

Asymmetrical Modular Multilevel Inverter with H-Bridge

When the DC sources of a CHB inverter are distinct, then an Asymmetric Cascaded H-bridge Multilevel Inverter is evolved. Binary MLI and trinary MLI are perhaps the most common ASMLI configurations. The ASMLI is categorized as ASMLI with and without H-bridge .

Panda *et al.* (2019) introduced the structure and control of a switched-diode MLI. The (Switched Diode Dual Source) SDDS MLI is initially developed with an asymmetric basis. The SDDS MLI proposed needs less number of switch and drivers than the few reduced switch MLI topologies recently developed. A modified version of the Fish Swarm Optimization (FSO) method is tested for the measurement of the best switching angles appropriate for the regulation of SDDS MLI to enhance the voltage efficiency by removing targeted low-order harmonics.

Eshwar Gowd *et al.* (2019) developed an 11-level asymmetric MLI with a 1:2 DC voltage ratio. A sliding mode control for the asymmetric multilevel inverter is also implemented. The author focused on the advantages of reduced switch asymmetric MLI topologies. This topology allows more efficient use of DC sources than the CHB topologies by using series-parallel switches.

Prem *et al.* (2019) proposed an asymmetric dual-source MLI which can generate 13-level of output wave form by means of 9 switches. The nearest level modulation strategy is handled for triggering the switches of the suggested inverter. The H-bridge is used for producing positive and negative levels.

Yahya *et al.* (2019) introduced a new Cascade Multilevel Inverter (CMLI) topology. The suggested topology proposes an optimal use of DC-source, minimized switch counting, and the removal of active switches in the direction to mitigate power losses. It can produce nearly twice the amount and can act as both symmetric and asymmetric inverter. Similar CMLI modules remove the need for a variety of semiconductors and render simple spare management. In addition, the suggested topology can effectively be applied to systems of high voltage.

Barzegarkhoo *et al.* (2018) proposed a novel single-stage switched capacitor multilevel inverter (SCMLIs). This offers a range of advantages over traditional topologies such as a suitable boosting effect, higher efficiency, lower DC voltage sources needed, and other components with lower complexity and cost. The basic structure of the proposed converter can produce a 9-level output voltage under various loading conditions. Two proposed switched capacitor cells are attached to obtain a new extended design for achieving a greater number of output voltage levels.

Ahamed Ibrahim *et al.* (2019) proposed a switched diode MLI. It contains a sequential connection of sub multilevel with H-bridge. The H- bridge is used for the production of positive and negative voltage levels. In symmetric mode the THD is 7.58% and in asymmetric mode the THD obtained is 7.701%.

Prabaharan & Palanisamy (2016) conferred a modular MLI. The basic unit of the suggested configuration consists of two switches and one DC source. The H-bridge circuit is needed for obtaining the positive and negative levels. The proposed configuration uses unipolar PWM-based pulse generation. In asymmetric mode operation, the Total Harmonic Distortion (THD) obtained is 4.25 % which is very less and meets the IEEE-519 requirements.

Asymmetrical Modular Multilevel Inverter without H-Bridge

Masoudinia *et al.* (2020) offered a novel structure of MLI, which produces 27-level of output voltage using 16 switches and 5 DC sources. The proposed structure is inherent (no H-bridge) in nature. Fundamental frequency switching strategy is implemented for firing the switches of the proposed MLI.

Majumdar *et al.* (2020) dealt with a multicell MLI. It consists of 8 unilateral switches, one bilateral switch and 3 DC sources to generate a 15- level of the output . It consists of a voltage tripler circuit. Nearest Level Control (NLC) modulation technique is implemented for switching the proposed inverter.

Siddique *et al.* (2019b) developed a MLI with reduced device count. The basic unit of the suggested configuration generates 13 output level with 3 DC power supplies and 8 power switches. A detailed description of the suggested structure to show the advantages of the suggested converter towards the other developed MLI topologies has been carried out. Power loss analysis was performed with Piecewise Linear Electrical Circuit Simulation (PLECS) software, resulting in 98.5 percent maximum efficiency. The NLC is operated to produce pulses for the switch to increase the voltage levels.

Siddique *et al.* (2019c) proposed a new single-phase cascaded MLI configuration. Three specific algorithms are suggested for evaluating the peak value of DC power supply. The aim of optimizing the suggested configuration is to achieve higher levels while reducing other parameters. A systematic assessment of the suggested structure is made with other comparable MLI-structures. The pulses to the switches are provided by a SHEPWM technique.

Siddique *et al.* (2019d) suggested two new topologies with a smaller number of switching criteria for producing the staircase output. The topology consists of 3 DC sources and 10 power switches to produce 15-level output. The second configuration consists of 4 DC sources and 12 switches to generates 25-level. These topologies demonstrate the benefits of decreased voltage stresses around the switches, except that they have lower switch counts.

Sandeep & Yaragatti (2018) improved a nine-level sensorless inverter topology with lower part numbers. It is created by a 3-level T-type NPC inverter with a two-level converter assembly fed by a Floating Capacitor (FC). In addition, the DC-link is paired with two line-frequency switches. A simple logic-figure equation-based pulse width modulator is created to preserve the FC voltage without the use of a voltage or current sensor at its reference value. Therefore, the difficulty of managing the proposed topology is very small.

Taheri & Samsami (2019) proposed a MLI created on the basis of switched-capacitor with limited DC power supply. The voltage production without any need for the H-bridge is the best feature of the suggested structure. In addition, the suggested topology has a flexible potential for extension to increase the output levels. The contrast of the suggested inverter with certain previous topologies indicates that suggested inverter has reduced component count.

Majareh *et al.* (2019) developed a MLI with reduced switch count. The provided MLI architecture is versatile and extensible. To assess the significance of DC voltage sources, three different approaches are implemented. Implementing SHEPWM strategy generates high quality of output voltage. The voltage produced is 9-level using symmetric DC power supplies. The use of asymmetric DC voltage sources produces 21 and 31-levels respectively.

Babadi *et al.* (2018) presented the problems associated with Packed U Cell and proposed two solutions. The first solution involves a detailed study of cascaded topologies for the existing basic units and the second method is a modern updated design focused on traditional converters to boost the PUC's output in terms of overall blocking voltage, switching ratings, and expanding its reliability to applications of high voltage.

Samadaei *et al.* (2018) introduced a square T-type MLI with reduced switch count. A unit contains 4 DC sources and 12 power switches. Each unit produces a voltage of 17- level. The units are cascaded to achieve further levels. For producing switching pulses, the nearest level control approach is used.

Arun & Noel (2018) proposed a cross-connected semi-half-bridge MLI. One half-bridge cell has a DC source, a diode, and a switch. Further semiconductor cells may be coupled in series to produce higher levels. By connecting the series strings of semi-half-bridge cells in a crisscross, the generalized structure is developed. With reduced switches and less blocking voltage, the suggested topology produces a staircase output. With an increase in output voltage levels, additional diodes are needed.

LITERATURE REVIEW OF MODULATION STRATEGIES

The modulation strategies are used to generate pulses for the switches in the power converter to produce the preferred voltage or current. All the modulation strategies are designed to generate a synthesized staircase wave shape of the voltage on the basis of the reference signal with various voltage and current amplitudes, frequency, and fundamental elements of the steady sinusoidal waveform.

In turn, these modulation strategies achieve a lower THD with better power quality and also retain controlled DC-link voltages and the RMS output voltage. The switches are triggered and disabled based on the pulses. These pulses are produced in various ways, and pulse production is categorized according to the comparison of standard and carrier signals. The modulation strategies are categorized on the basis of the switching frequency.

The overall modulation methods specifications are as follows:

- Lower switching frequency.

- Lower output total harmonic distortion.

- Equal power sharing among the power devices.

- Effective control method.

- Implementation costs should be minimal.

Prabaharan *et al.* (2020) proposed a multicarrier unipolar pulse width method to generate switching pulses to the multilevel inverter (MLI) which uses DC sources of trinary series. It offers the highest degree of output voltage with a minimum DC source and switching count, as opposed to other series, such as symmetrical, normal, binary, and quasi-linear. The multicarrier unipolar pulse width method triggers the 9-level asymmetric inverter and produces output voltage with less THD.

Basante *et al.* (2020) implemented the SHE-PWM technique to control the circulating current in the modular multilevel converter. The control technique adjusts the DC component of the circulating current and the stored energy to its reference value. Seventeen switching angles are generated by the PWM technique to optimize the circulating current and harmonics.

Hosseinpour *et al.* (2020) presented a 7-level reduced switch multilevel inverter. The author implemented the SPWM and SHE PWM techniques to the proposed structure. By implementing the SPWM technique and SHE PWM technique the THD obtained is 10.02% and 8.35% .From the results obtained it is evident that the SHE PWM technique reduces the harmonics in the system.

Janabi *et al.* (2020) presented a new solution to substantially limit the harmonics using the SHE PWM technique. The transcendental optimization equations are translated into algebraic equations with Chebyshev polynomials. The approach proposed is to generalize the Chudnovsky algorithm, requiring rather than removing the modulation of the chosen harmonics. In comparison, the method used to extract the roots of the polynomial enables the algorithm to be applied in real-time applications.

Haghdar (2020) presented the SHE optimization approach in CHB MLIs using Teaching-Learning-Based Optimization (TLBOs). The main aim of SHE is to remove harmonic contents by computing the nonlinear equations and the equations are computed using the TLBO algorithm. The third and fifth harmonics are totally removed from the output using the SHE-TLBO technique.

De Simone *et al.* (2020) presented the windowed modulation method for MMC in traction systems. The windowed modulation is developed from NLC PWM, in order to incorporate its intrinsic capabilities with increased flexibility. It eliminates the disadvantages of switching in contrast to traditional PWM devices and decreases existing harmonic fluctuations in conjunction with NLC schemes.

LITERATURE REVIEW ON OPTIMIZATION TECHNIQUES

Priyadarshi *et al.* (2020) introduced a hybrid (ANFIS–PSO) hybrid (maximum power point tracking) MPPT method of adaptive neuro-fuzzy inference method for gaining fast and full PV capacity with zero oscillating monitoring. The technique used provides remarkable power to boost PV extraction

Routray *et al.* (2020) proposed a modified grey wolf optimization for a 11-level inverter. SHE PWM technique is used to compute the firing angles and the harmonic ripples are minimized.

Falehi (2020) suggested a new asymmetric multilevel inverter with the Multi Objective Bee Algorithm (MOBA). The suggested SHE PWM technique eliminates the lower order harmonics.

Salman *et al.* (2020) used the GA based SHE PWM technique in CHBMLI. Selective Harmonic Elimination (SHE) is the most common and professional strategy for low-order harmonic minimization but requires nonlinear transcendental equations which are quite challenging to solve analytically and numerically. For the 5-level inverter, two switching angles were calculated and the harmonics were reduced.

Qais *et al.* (2020) proposed the Harris Hawks Optimization (HHO) method for evaluating the optimization constraints. The optimization is based on the hunting style of the harris hawks. These hawks attack the prey by using certain methods. Results prove that this algorithm is very fast in solving the problems.

Shaik & Ganesh (2019) presented a firefly technique for training (Artificial Neural Network) ANN to restore the power from a voltage source converter during a blackout. The drawback of the soft start-up has been overcome by implementing the ANN-FLM method. A comparison analysis is done in between the ANN-FLM and conventional PI controller and the outcomes prove that ANN-FLM controller is dynamic.

Subramaniyan & Ramiah (2020) implemented the improved football game optimization for (Hybrid Active Power Filter) HAPF. The usage of advanced football algorithms in the erection of HAPF shows the supremacy of algorithms in delivering the best global architecture, reducing overall harmonic distortion, and increasing power factor at the lowest possible amount.

LITERATURE REVIEW ON GRID CONNECTED SYSTEMS

Luo *et al.* (2020) proposed the asymmetric NPC for a single-phase grid connected system. The authors implemented a modified single carrier based SPWM technique for reducing the harmonics in the injected grid current. A Proportional Resonant (PR) controller is implemented for closed loop control with enhanced phase lock loop. The THD of the grid current is obtained as 4.9% which satisfies the IEEE 519 standards.

Gude & Chu (2020) proposed a Multiple Delayed Signal Cancellation Filter (MDSCF) in the PLL to reduce the harmonic distortions in the phase angle and amplitude determination loop. Further the transient response of the system is examined by implementing the enhanced PLL with Propotional Integrative Derivative (PID) controller.

Vosoughi *et al.* (2020) proposed a transformer less switched-capacitor based grid-connected inverter. It contains 6 unilateral switches and 2 DC power sources to produce 5-level of the output voltage. Perturb and Observer algorithm harvests the utmost power from PV panels. A peak current regulator is used to regulate current of the inverter for Grid Connected System (GCS).

Dutta & Chatterjee (2020) introduced the coupled inverter type buck-boost converter for a single-phase grid connected PV system. The main aim of this converter is to reduce the PV modules in a series connected sub array. An inductor current control technique is implemented to inject harmonic less power in to the grid.

Makhamreh *et al.* (2020) proposed the Sliding Mode Control (SMC) for packed U type grid-connected MLI. The control complexity in PUC converter solved by using the sliding mode control. The grid connected system is effectively analysed using the SMC and (Modern Predictive Controller) MPC. The grid current has a THD of 0.96% using the SMC controller while the MPC controller gives a THD of 0.75%.

Alluhaybi *et al.* (2020) presented a review on single phase grid connected micro inverters. Various topologies are compared on the basis of cost, switching techniques, device count and range of input voltage. The voltage boosting techniques in the PV system are obtained using single or double stage DC-DC converters. The dominant and efficient topology is simulated and analysed.

Bana *et al.* (2020) proposed a Voltage Level Boost (VLB) asymmetric 15-level inverter for the grid connected system. The switching pulses are provided using PDPWM scheme. An incremental conductance technique harvests utmost power from the PV panels. For voltage regulation in the grid connected system a PI controller is used.

Alqatamin & Mcintyre (2020) proposed a filter-based control method for the single-phase grid connected PV system. A cascaded Proportional Resonant (PR) controller is also implemented and compared with the proposed controller. The superiority of the proposed controller is analysed by measuring the THD of the grid current. The filter-based controller thus proves the best by producing a current THD of 2.5% over the PR controller which produces a current THD of 5.84%.

Mohapatra & Agarwal (2020) proposed an improved model predictive controller for a single-phase grid connected system. A 5-level NPC is used as the voltage source inverter in grid connected system. The suggested control method improves the current harmonics in grid connected system.

Kanavaros *et al.* (2020) implemented the mixed frame control method for the grid connected PV system. Different methods to control the active and reactive power with orthogonal generation methods were analysed. The proposed mixed frame control method reduces the harmonics in grid current and controls the active and reactive power.

SUMMARY

The basic principles and various kinds of modulation techniques and MLI structures are discussed. Based on the literature it is found that all the works are carried for low voltage systems. Moreover, some inverters have more switches and a greater content of THD. Table 2.1 presents the literature on MMI ,based on the voltage levels, switch count,PWM technique and load.

Table 2.1 Literature Review of MMI

Authors	Year	Switches	Level	Techniques	THD	Load
Symmetrical Modular Multilevel Inverter with H-Bridge						
Dewangan *et al.*	2019	7	9	MCPWM	14.06%	RL
Gandomi *et al.*	2019	12	13	PWM	9.16%	RL
Siddique *et al.*	2019a	11	9	NLCPWM	9.4%	RL
Sandeep *et al.*	2019	9	5	MCPWM	———	RL
Salehahari *et al.*	2019	8	9	MCPWM	———	RL

Authors	Year	Switches	Level	Techniques	THD
Symmetrical Modular Multilevel Inverter with H-Bridge					
Lee *et al.*	2018a	10	9	SHEPWM	———
He & Cheng	2018	8	7	SPWM	11.20%
Panda & Panda	2018	6	7	SHE-PSO	11.81%
Lee	2018b	16	7	LSPWM	———
Symmetrical Modular Multilevel Inverter without H-Bridge					
Siwakoti *et al.*	2020	8	7	LSPWM	2.16%
Liu *et al.*	2020	9	7	SPWM	12.23%
Mahato *et al.*	2020	10	9	LSPWM	13.63%
Roy & Sadhu	2020	14	13	FSF-PWM	7.45%
Lee & Lee	2020	10	7	SPWM	———
Bhatnagar *et al.*	2019	11	9	LSPWM	———

Bhatnagar *et al.*	2019	8	9		LSPWM	———	RL
		10			SHE-GA	19.05%	RL
			7				
		10			SPWM	18.64%	RL
Lee *et al.*	2018c						
		16			SHE-GA	8.34%	RL
			13				
		16			SPWM	9.33%	RL

Asymmetrical Modular Multilevel Inverter with H-Bridge

Panda *et al.*	2019	6	7	SHE-FSO	10.51%	RL
Eshwar Gowd *et al.*	2019	12	11	PWM	11.93%	RL
Prem *et al.*	2019	9	13	NLCPWM	6.25%	RL
Yahya *et al.*	2019	10	11	PWM	———	RL
Barzegarkhoo *et al.*	2019	10	49	PWM	———	RL

Authors	Year	Switches	Level	Techniques	THD	Load
Asymmetrical Modular Multilevel Inverter with H-Bridge						
Ahamed Ibrahim et al.	2019	9	11	NLCPWM	7.58%	RL
		18	13	NLCPWM	7.701	RL
Prabaharan & Palaniswamy	2016	12	31	MCPWM	4.25%	RL
Asymmetrical Modular Multilevel Inverter without H-Bridge						
Masoudinia et al.	2020	16	27	FSF-PWM	———	RL
Majumdar et al.	2020	9	15	NLCPWM	5.66%	RL
Siddique et al.	2019b	8	13	NLCPWM	6.30%	RL
Siddique et al.	2019c	8	11	SHEPWM	———	RL
Siddique et al.	2019d	10	15	NLCPWM	———	RL
		12	25	NLCPWM	———	RL
Sandeep & Yaragatti	2018	8	9	FSF-PWM	———	RL

Taheri & Samsami	2019	12	17	PWM	———	RL
		10	9	FSF-PWM	5.58%	RL
Majareh *et al.*	2019	10	21	FSF-PWM	3.95%	RL
		10	31	FSF-PWM	2.79%	RL
		22	49	FSF-PWM	1.63%	RL
Babadi *et al.*	2019	66	147	FSF-PWM	0.72%	RL
Samadaei *et al.*	2018	12	17	NLCPWM	3.06%	RL
		10	9	MCPWM	13.45%	RL
Arun & Noel	2018	10	21	MCPWM	5.67%	RL

CHAPTER 3

DESIGN AND DEVELOPMENT OF SYMMETRICAL AND ASYMMETRICAL MODULAR MULTILEVEL INVERTER

INTRODUCTION

MLIs are designed primarily for their output, which is a near sinusoidal waveform. These inverters are used in potential applications of high voltage with low voltage rated switches (Barbie *et al.* 2020). MLIs generate AC voltage using distinct DC power supply. These DC power input may be symmetric or asymmetric in nature. Technologically, the multilevel inverter can be used for various applications. MLIs can achieve high voltages and remove harmonic distortions without using transformers when connected to grid. The DC voltages can be converted to the AC output, which will lead to a near sinusoidal waveform with low distortions.

Conventional PWM inverters need a carrier signal more than 1kHz to obtain an output waveform of 50Hz frequency. But MLIs mostly use a fundamental switching frequency of 50Hz to obtain the output. This arrangement will, therefore, lead to a minimal changeover loss equivalent to other PWM schemes. As a result, along with traditional configurations, researchers are still developing new structures with an application-oriented approach.

The work aims at appreciating developments that take place in many forms and continue to be a hot trend. In addition, the development of

Modular Multilevel Inverter (MMI) is an important achievement to the development of MLI configurations. For medium or high-power implementations, it has become an attractive topology. A limited number of components are added in order to address these challenges. Even though many structures have so far been developed based on the reduced number of components, some structures are still in progress. In general, three modalities accompany the development of the new topology: structural improvement; the introduction of the unequal DC source as an alternative of the equal DC source, the amalgamation of structural improvement, and placement of the asymmetric DC source.

PROBLEM IDENTIFICATION

The implementation of MMI is much essential to maintain a good quality of power in the power system. Due to the nonlinear characteristics of the traditional inverters the output power cannot meet the IEEE grid code specifications. Therefore, LC filters are applied on the output side of the traditional inverters to remove harmonics and progress the power quality. While stress on the switches is typically not minimized, the system's reliability and performance are decreased (Sathik & Vijayakumar 2019). The issues of the traditional inverters can be solved by using multilevel inverters. Due to the numerous advantages, MLIs are much favoured in numerous applications. The use of high frequency modulation approaches produces voltage stress and reduces the inverter output.

However, when the output tends to increase, the switching elements also increase in a conventional MLI. As the amount of switching components increases the associated gate drivers also increase. Accordingly, switching the components become difficult and affect the efficiency of the inverter. Moreover, the losses of the inverter are also increases. The construction of reduced device count MLIs is additionally a practice aimed at reducing both production costs and energy costs indirectly due to on-off losses and related heat sinks. When assessed with the conventional MLIs, the reduced part count configurations greatly decrease the switch count because of implementing the switches in different patterns.

Many modulation methods like SPWM, SHEPWM, MCPWM, Digital PWM are used in reduced switch topologies. The big disadvantage of the reduced MLI systems is that they cannot completely utilize the benefits of minimized switches owing to the large frequency switching errors in some schemes. The MLI-HDBK-2017F (Najafi & Yatim 2012) standard cites that increased device count affects the reliability of the inverter. In any power electronic system, the system efficiency is inversely correlated with the device part count. Reducing the devices increases the stability and also boost system performance. Therefore, power electronic systems with reduced component count improve the durability and effectiveness of the system.

DESIGN OF MODULAR MULTILEVEL INVERTER

A large number of electronic components are needed in most of the present MLIs. A new modular MLI is implemented to reduce power electronic components. This MMI topology consists of 6 unidirectional switches, 4 bidirectional switches, and 5 DC sources. These elements are arranged intelligently to produce the required output. The bidirectional switches are also known as two way switches. These switches are mostly used in power conversion systems.

The function of the bidirectional switch is to block the positive or negative voltages as well as allow the flow of currents in any direction. These two way switches are triggered on and off in a controllable way so that the require output voltage and current can be obtained. These switches are mostly used in the design of MLIs, Voltage Source Inverter (VSI), Matrix converter and voltage regulators. Basically, a two way switch has two major components one is the semiconductor switch and the other is its driver circuit. The (Metal Oxide Field Effect Transistor) MOSFET is used in designing the bidirectional switches for low power and high switching frequency applications.

The Integrated Gate-Commutated Thyristor (IGCT), Gate Turn-Off thyristor (GTO) and MOS Turn-Off (MTO) thyristor are used in designing the bidirectional switches for high power applications but with limited switching frequency. For high power and high switching frequency applications the Insulated-Gate Bipolar Transistors (IGBTs), Silicon Carbide MOSFETs (SiC- MOSFETs) and Silicon MOSFETs (Si-MOSFETs) are used. The major benefits of implementing bidirectional switches are low switching losses, high temperature resistant and capable of operating at high voltage and switching frequencies.

There are four different types of bidirectional switches namely the Diode Bridge type (DB), Common Emitter (CE) type, Common Collector (CC) type and the Reverse Blocking-IGBT (RB-IGBT) type (Maqueda *et al.*2018) . The diode bridge type bidirectional switch is the first developed topology. This structure consists of four diodes in a bridge configuration and an IGBT in the middle as shown in Figure 3.1 a. This structure produces more losses since three devices are in conduction all the time.

The common emitter configuration shown in Figure 3.1 b consists of two IGBTs and two diodes. These elements are connected in anti-parallel and anti-series. The diodes are implemented for performing reverse blocking of voltage. The direction of current can be controlled separately using this configuration. Since only two devices are conducting at an instant the conduction losses are minimum in this configuration. Moreover, the CE configuration needs only a single driver circuit for operation (Sadanala *et al.* 2020).

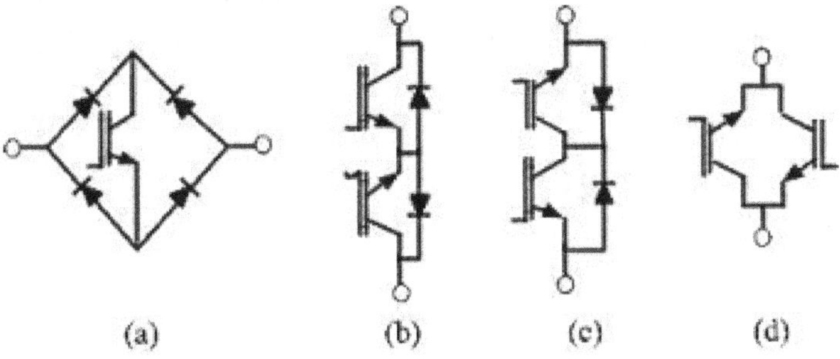

Figure 3.1 Structure of Bidirectional Switches (a) Diode Bridge (DB)

**(b) Common Emitter (CE) (c) Common Collector (CC)
(d) Reverse Blocking IGBT RB-IGBT**

The common Collector configuration (Sebaaly *et al.* 2020) is shown in Figure 3.1c. It also consist of same number of devices like the CE configuration. The only drawback of the CC configuration is that it cannot be implemented for large systems since stray inductances cause triggering issues. The RB-IGBT is shown in Figure 3.1d. It consists of two RB-IGBTs connected in anti-parallel. This topology causes reduced voltage drop in the switch due to the absence of diode since the current flows to the load through only one single switch.

In the proposed work the common emitter configuration is used since it has fewer conduction losses and needs a single gate driver circuit. The suggested configuration is suitable for solar PV applications since 5 isolated DC sources can be used. In order to achieve the negative voltage levels, the fifth DC source is activated by triggering the P_2 switch. The value of the fifth

DC source is chosen as the sum of the four individual DC sources. Therefore, H-bridges are not necessary for obtaining the negative level.

The basic structure and the basic unit of MMI is presented in Figure 3.2. It consists of four PV sources and the fifth source (V_{peak}) is obtained by connecting the four PV sources in series. The PV panels are fed with DC-DC converters to obtain constant voltage at the output. Figure 3.2b presents the basic unit of the proposed MMI. The basic unit consist of a PV source, three unidirectional switches (S, P, N), a bidirectional switch (B), a peak voltage source (V_{peak}), and the RL load.

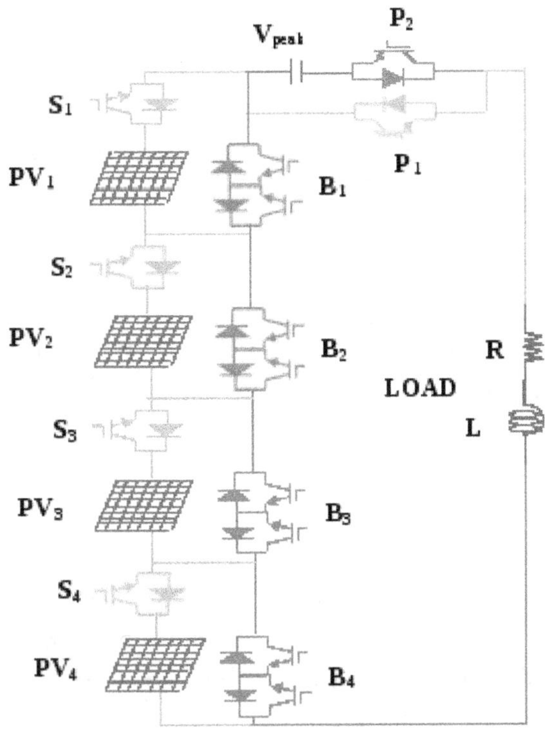

(a)

Figure 3.2 (Continued)

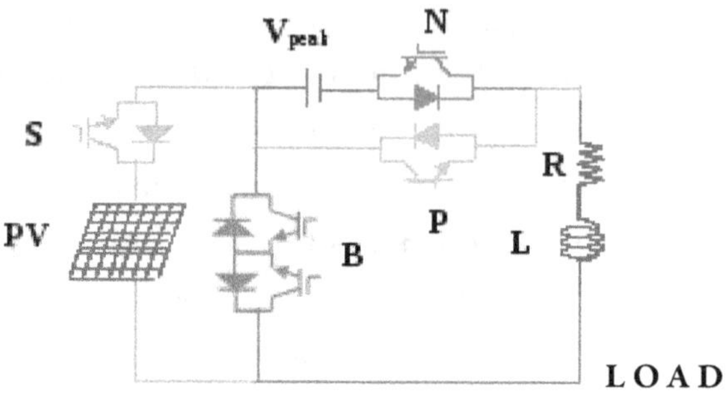

(b)

Figure 3.2 (a) Basic structure of the MMI (b) Basic unit of MMI

In the basic circuit when switch S is triggered along with switch P the positive voltage level (+PV) is generated. For obtaining the negative voltage level the switches N and B must be triggered. When switch N is triggered the peak voltage is activated and is passed through the bidirectional switch (B) with a negative polarity thus the negative voltage level (-PV) is generated. The peak voltage of the basic unit is same as the PV voltage and this condition is only suitable for Figure 3.2b. This peak voltage value changes when additional units are added. The zeroth level is generated by switching the unidirectional switch P and the bidirectional switch B. Therefore, the basic unit generates a three level of output voltage (+PV,0V, - PV).

Adding an additional unit to the basic unit can generate a five level of output voltage. If we consider two units where there are two PV sources (PV_1,PV_2), then the peak voltage or the fifth DC voltage value will be the sum of the PV sources (PV_1+PV_2). Likewise if we take the Figure 3.2a the peak voltage of the basic structure is PV_1+ PV_2+PV_3+PV_4 .Therefore by triggering the unidirectional switches (S_1, S_2, S_3, S_4) along with the switch P_1 the positive voltage levels are obtained and by triggering the unidirectional switches

(S_1, S_2, S_3, S_4) along with the switch P_2 the peak voltage(V_{peak}) gets activated and the negative voltage levels are obtained.

The proposed structure can be extended to several levels. The cascaded connection of the suggested structure is shown in Figure 3.3. In the basic structure if symmetric voltages are implemented then a 9-level output voltage is generated. If two such modules are connected in a cascaded form then a 17-level output is generated.

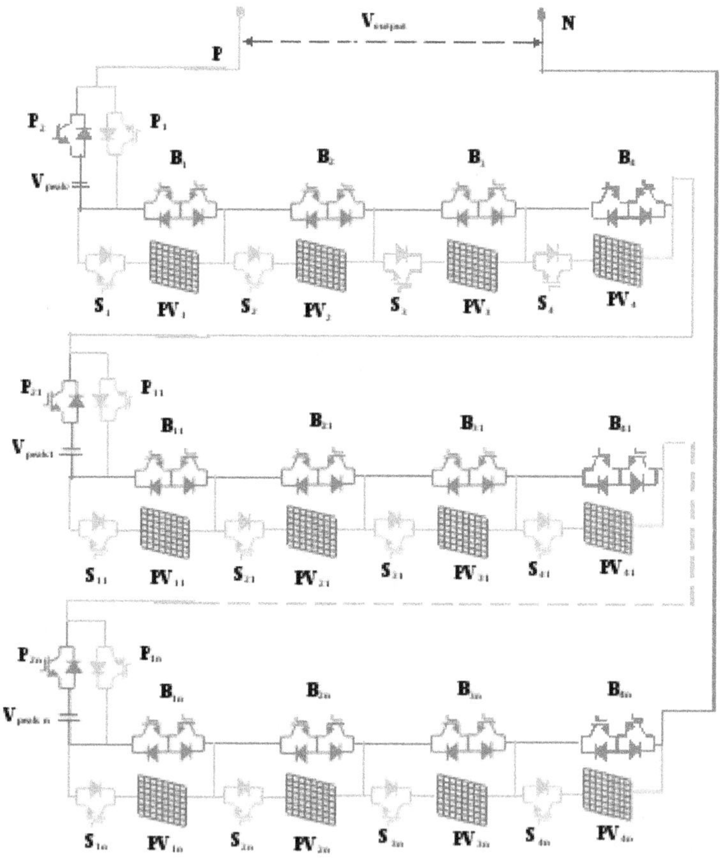

Figure 3.3 Cascaded structure of MMI

Therefore, the generalised formulae for the cascaded connection is

$$N_{LC} = (2V_{peak} + 1)n - (n - 1) \quad (3.1)$$

Where N_{LC} is the number of levels obtained by cascading the

connection, 'n' is the number of modules and V_{peak} is the peak voltage of the

basic unit.

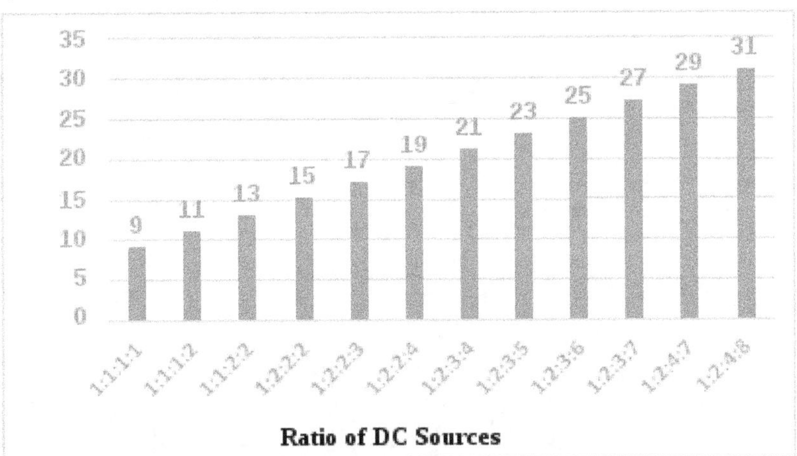

Figure 3.4 illustrates the graphical presentation of possible voltage levels obtained from the proposed inverter using symmetric and asymmetric DC sources.

Figure 3.4 Possible voltage levels obtained using symmetric and asymmetric DC sources

SYMMETRIC MODULAR MULTILEVEL INVERTER TOPOLOGY

When comparing with the three conventional MLI configurations, the CHB MLI has special qualities that are capable of using either equal or unequal magnitude of DC source voltage to achieve the required output. The

CHB structures which use equal and unequal DC sources are called symmetric and asymmetric configurations. In this way, the proposed MMI also can be constructed in both symmetric and asymmetric configurations. The suggested inverter can be used as SMMI as well as AMMI depending on the input DC power supply.

The DC sources should be in 1: 1: 1: 1 ratio for obtaining a

symmetric 9-level MMI. Figure 3.5 and 3.6 illustrate the modes of operation of the proposed symmetric MMI with its output voltage waveforms. The fifth

DC source (V_{peak}) is chosen by Equation 3.2.

$$V_{peak} = \frac{N_L - 1}{2} \quad \text{or} \quad V_{peak} = L \, V_{PV1\text{-}4}$$

$$(3.2)$$

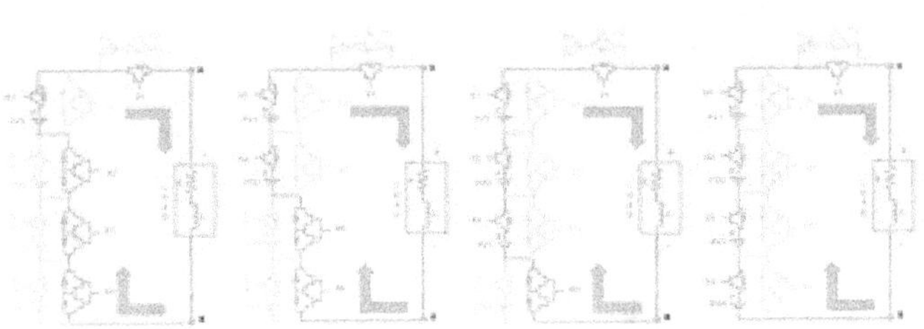

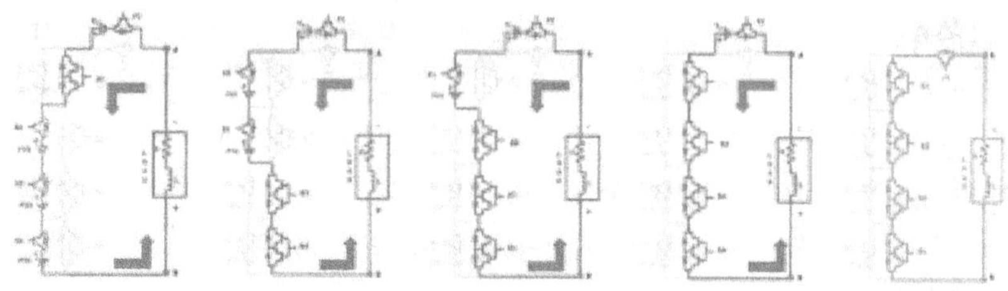

1V_{DC} 2V_{DC} 3V_{DC} 4V_{DC}

-1 V_{DC} -2 V_{DC} -3 V_{DC} -4 V_{DC} OV

Figure 3.5 Modes of Operation of Symmetric 9-level MMI

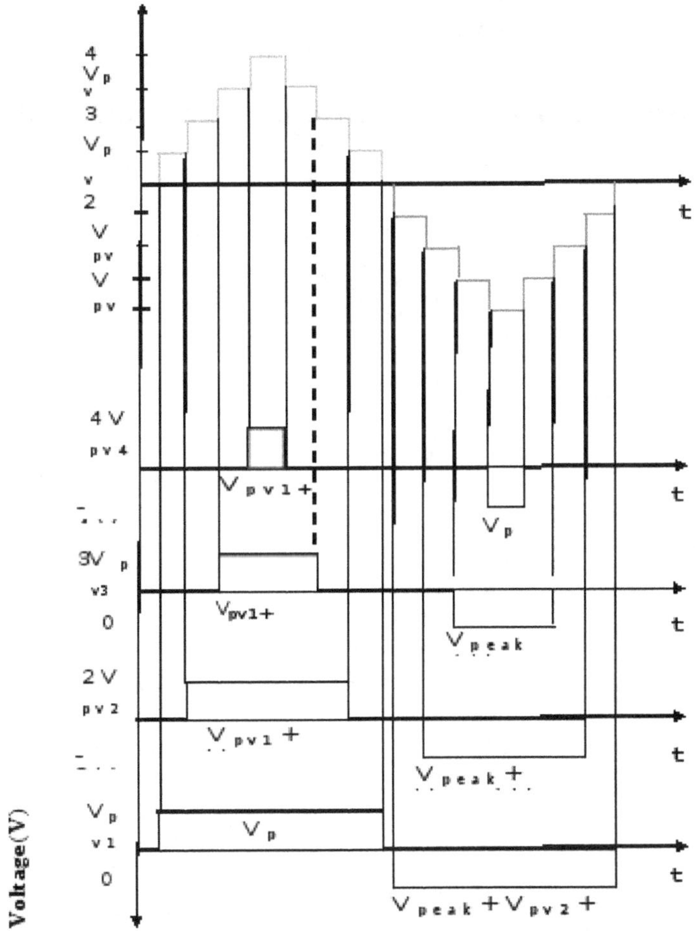

Where, N_L is the number of levels obtained, V_{PV} is the voltage from PV panels.

Note : V_{peak} is taken as $(-4V_{pv})$

Figure 3.6 9-level output of the proposed Symmetric MMI

Table 3.1 indicates the strategy of switching for the suggested symmetric MMI. Table 3.2 specifies the parameters of the symmetric MMI. Table 3.3 presents the current path of the proposed symmetric MMI.

Table 3. 1 Switching Sequence of the Symmetric MMI

S1	S2	S3	S4	B1	B2	B3	B4	P1	P2	Stages of operation	Output Voltage (V_o)
✓	Ξ	Ξ	Ξ	Ξ	✓	✓	✓	✓	Ξ	I	1 V_{DC}
✓	✓	Ξ	Ξ	Ξ	Ξ	✓	✓	✓	Ξ	II	2 V_{DC}
✓	✓	✓	Ξ	Ξ	Ξ	X	✓	✓	Ξ	III	3 V_{DC}
✓	✓	✓	✓	Ξ	Ξ	X	X	✓	Ξ	IV	4 V_{DC}
Ξ	Ξ	Ξ	Ξ	✓	✓	✓	✓	✓	Ξ	XIV	0 V_{DC}
Ξ	✓	✓	✓	✓	Ξ	Ξ	Ξ	X	✓	XV	-1V_{DC}
✓	✓	Ξ	Ξ	Ξ	Ξ	✓	✓	X	✓	XVI	-2V_{DC}
✓	Ξ	Ξ	Ξ	Ξ	✓	✓	✓	X	✓	XVII	-3V_{DC}
Ξ	Ξ	Ξ	Ξ	✓	✓	✓	✓	X	✓	XVIII	-4V_{DC}

✓ =ON, Ξ =OFF

Table 3.2 Parameters of the Symmetric MMI

Parameters	Based on Number of Levels	Based on Number of Modules
Number of levels	N_L	$9n$
Number of switches	$(N_L + 5)$	$9n + 5$
Number of diodes	$(N_L + 5)$	$9n + 5$
Number of drive circuit	$(N_L + 5)$	$10n$
Number of DC sources	$(N_L - 4)$	$5n$
Total Standing Voltage (TSV)	$(2N_L - 1)$	$17n$

Table 3.3 Current Path for 9-Level MMI

Output voltage	Switches in conduction	Current path
$+Vdc$	S1, P1, B2, B3, B4	PV_{1+} S1 P1 Load B4 B3 B2 PV_{1-}
$+2Vdc$	S1, S2, P1, B3, B4	PV_{2+} S2 PV_1 S1 P1 Load B4 B3 PV_{2-}
$+3Vdc$	S1, S2, S3, P1, B4	PV_{3+} S3 PV_2 S2 PV_1 S1 P1 Load B4 PV_{3-}
$+4Vdc$	S1, S2, S3, S4, P1	PV_{4+} S4 PV_3 S3 PV_2 S2 PV_1 S1 P1 Load PV_{4-}
$0Vdc$	B1, B2, B3, B4, P1	B1 P1 Load B4 B3 B2 B1
$-Vdc$	P2, S1, S2, S3, B4	V_{peak+} S1 PV_1 S2 PV_2 S3 PV_3 B4 Load P2 V_{peak-}
$-2Vdc$	P2, S1, S2, B3, B4	V_{peak+} S1 PV_1 S2 PV_2 B3 B4 Load P2 V_{peak-}
$-3Vdc$	P2, S1, B2, B3, B4	V_{peak+} S1 PV_1 B2 B3 B4 Load P2 V_{peak-}

$-4V_{dc}$	P2, B1, B2, B3, B4	V_{peak+} B1 B2 B3 B4 Load P2 V_{peak-}

COMPARISON OF THE PROPOSED 9-LEVEL SYMMETRIC MMI WITH CONVENTIONAL 9-LEVEL INVERTER TOPOLOGIES

To validate the suggested structure on the basis of switch count, the MMI should be compared with other MLI configurations. Table 3.4 presents the comparison of the proposed symmetric MMI configuration with other 9-level symmetrical configurations.

Table 3.4 Comparison of the suggested Symmetric MMI with conventional 9-Level Symmetrical Topologies

Topology	N_L	N_{DC}	N_{sW}	N_C	N_D	Negative levels
NPC	9	1	16	0	56	Two arm
FC	9	1	16	28	0	Two arm
CHB	9	4	16	0	16	H-bridge
Veenstra & Rufer (2005)	9	2	14	2	14	H-bridge
Chaudhuri & Rufer (2010a)	9	2	14	2	14	Inherent
Chaudhuri *et al.* (2010b)	9	1	14	4	14	Inherent
Babaei (2008)	9	4	20	0	20	H-bridge
Li *et al.* (2010)	9	2	12	3	12	H-bridge
Jun Li *et al.* (2011)	9	1	14	4	14	Inherent
Rajeevan *et al.* (2012)	9	2	12	3	12	H-bridge
Kangarlu & Babaei (2013)	9	4	12	0	12	H-bridge

Nagarajan & Saravanan (2014)	9	4	11	0	11	H-bridge
Prabaharan & Palanisamy (2016)	9	4	12	0	12	H-bridge
Jahan *et al.* (2019)	9	1	17	4	21	H-bridge
Proposed	9	5	14	0	0	Inherent

From Table 3.4 it is observed that some configurations have the limitation of using large quantity of capacitors or power diodes and some topologies use H-bridges for producing negative levels. The power diode dissipates a lot of voltage during current conduction. When the diode is switched off, high voltage spikes arise in diode because of the stray inductance. The increasing proportion of faults in most power converters is defined by the capacitor.

Capacitor's failure rate can be assessed, by its operating conditions. Due to which more care is taken to safeguard the power diodes, capacitors from open circuit and short circuit faults. The use of H-bridges for producing negative levels causes losses in the inverter and the switches of the H-bridge are extremely prone to faults. An additional snubber circuit is needed for the protection of the switches of the H- bridge. The suggested inverter requires less switches and no additional diodes, capacitors or H-bridge are used in the suggested structure.

ASYMMETRIC MODULAR MULTILEVEL INVERTER TOPOLOGY

3.6.1 General Description

The above-mentioned symmetric topology uses more switches for producing higher output levels, which lead to an increase in switching complexity, gate driver circuits, cooling system, installation size, and complexity in the switching pulse generation. To address these limitations, the asymmetric configuration is highly preferred. The asymmetric configuration uses the same quantity of switches and generates various voltage levels. Using different ratios of asymmetric DC sources, the same symmetric structure can produce 11,13,15,17,19,21,23,25,27,29 and 31- levels of output.

A 27-level Asymmetric Modular Multilevel Inverter (AMMI) is produced in the same symmetric structure when the input DC power supply is

1: 2: 3: 7: V_p. There are 14 switches in the suggested structure of which 6 are one-way switches and 4 are two-way switches. The suggested structure contains 5 input DC sources. The fifth input DC source (V_P) is responsible for producing the negative levels in the suggested structure. Therefore, V_p value is selected as the total magnitude of all the four independent DC inputs. In the suggested 27-level inverter the V_p is 13V by considering $(1V + 2V + 3V +$

$7V = 13V)$.Table 3.5 indicates the triggering method for the 27-level AMMI

The power switches are intelligently designed to get the possible voltage levels. Thirteen positive, thirteen negative and zero level combine to form a 27-level. The positive thirteen level is generated by triggering the switches S1 to S4, B1 to B4 randomly along with switch P1. To generate the negative thirteen level in the suggested configuration, the switches S1 to S4, B1 to B4 are randomly switched along with switch P2. By switching all the bidirectional switches along with switch P1 the zeroth level is produced. Table 3.6 presents the parameters of the suggested topology.

The parameters presented in Table 3.6 is designed by considering the number of levels(N_L) and sectional units or modules (n). Higher voltage

levels can be generated by the suggested configuration when connected in cascaded form. The cascaded structure is illustrated in Figure 3.3. When two 27-level AMMI structures are connected in the cascaded manner a 53-level (26 positive level,26 negative level and zero level) is generated at the output. The modes of operation of the suggested 27-level AMMI is displayed in Figure 3.7 and the respective current path is shown in Table 3.7. The triggering design of the suggested configuration for a full cycle is presented in Figure 3.8. The suggested system can be expanded to multiple units. In the suggested structure, when the input DC sources are altered there is a modification in the output level.

Eleven different output voltage levels can be generated by the suggested AMMI using 11 different configurations of DC sources. Table 3.8 presents the possible outputs produced by the suggested AMMI using 11 different configurations.

Table 3.5 Firing Sequence of the 27-Level AMMI

S1	S2	S3	S4	B1	B2	B3	B4	P1	P2	Stages of operation	Output voltage (V_o)
✓	Ξ	Ξ	Ξ	Ξ	✓	✓	✓	✓	Ξ	I	1 Vdc
Ξ	✓	Ξ	Ξ	✓	Ξ	✓	✓	✓	Ξ	II	2 Vdc
Ξ	Ξ	✓	Ξ	✓	✓	X	✓	✓	Ξ	III	3 Vdc
✓	Ξ	✓	Ξ	Ξ	✓	X	✓	✓	Ξ	IV	4 Vdc
Ξ	✓	✓	Ξ	✓	X	X	✓	✓	Ξ	V	5 Vdc
✓	✓	✓	Ξ	Ξ	X	X	✓	✓	Ξ	VI	6 Vdc
Ξ	Ξ	Ξ	✓	✓	✓	✓	X	✓	Ξ	VII	7 Vdc
✓	Ξ	Ξ	✓	X	✓	✓	X	✓	Ξ	VIII	8 Vdc
Ξ	✓	Ξ	✓	✓	Ξ	✓	X	✓	Ξ	IX	9 Vdc
Ξ	Ξ	✓	✓	✓	✓	X	X	✓	Ξ	X	10 Vdc
✓	Ξ	✓	✓	X	✓	X	X	✓	Ξ	XI	11 Vdc
Ξ	✓	✓	✓	✓	X	X	X	✓	Ξ	XII	12 Vdc

✓	✓	✓	✓	X	X	X	X	✓	Ξ	XIII	13 Vdc
Ξ	Ξ	Ξ	Ξ	✓	✓	✓	✓	✓	Ξ	XIV	0 Vdc
Ξ	✓	✓	✓	Ξ	✓	✓	✓	X	✓	XV	-1 Vdc
✓	Ξ	✓	✓	✓	Ξ	✓	✓	X	✓	XVI	-2 Vdc
Ξ	Ξ	✓	✓	Ξ	Ξ	✓	✓	X	✓	XVII	-3 Vdc
Ξ	✓	Ξ	✓	Ξ	Ξ	✓	✓	X	✓	XVIII	-4 Vdc
✓	Ξ	Ξ	✓	✓	✓	✓	✓	X	✓	XIX	-5 Vdc
Ξ	Ξ	Ξ	✓	Ξ	✓	✓	✓	X	✓	XX	-6 Vdc
✓	✓	✓	Ξ	Ξ	✓	✓	✓	X	✓	XXI	-7 Vdc
Ξ	✓	✓	Ξ	✓	✓	✓	✓	X	✓	XXII	-8 Vdc
✓	Ξ	✓	Ξ	Ξ	Ξ	✓	✓	X	✓	XXIII	-9 Vdc
Ξ	Ξ	✓	Ξ	✓	Ξ	✓	✓	X	✓	XXIV	-10 Vdc
Ξ	✓	Ξ	Ξ	✓	Ξ	✓	✓	X	✓	XXV	-11 Vdc

| ✓ | Ξ | Ξ | Ξ | Ξ | ✓ | ✓ | ✓ | X | ✓ | XXVI | -12Vdc |
| Ξ | Ξ | Ξ | Ξ | ✓ | ✓ | ✓ | ✓ | X | ✓ | XVII | -13Vdc |

✓ =ON, Ξ=OFF

Table 3.6 Parameters of the 27-Level AMMI

Parameters	Based on Number of Levels	Based on Number of Modules
Number of levels	N_L	$26n+1$
Number of switches	$\dfrac{N_L+1}{2}$	$13n+1$
Number of diodes	$\dfrac{N_L+1}{2}$	$13n+1$
Number of Drive circuit	$\dfrac{N_L-7}{2}$	$10n$
Number of DC sources	$\dfrac{N_L-7}{4}$	$5n$
Total Standing Voltage (TSV)	$5\left[\dfrac{N_L-1}{2}\right]$	$65n$

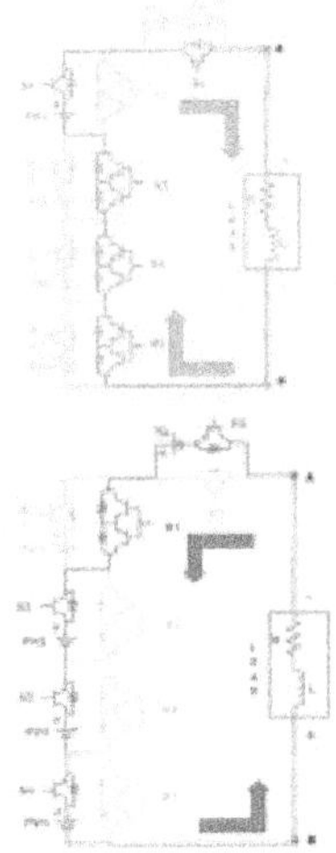

$$V_{DC} \; -V_{DC} \; 2V_{DC} \; -2V_{DC}$$

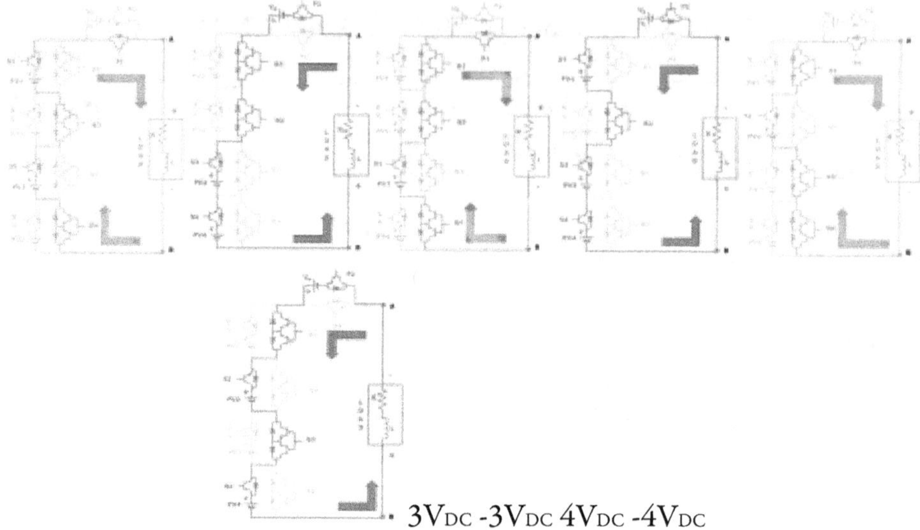

Figure 3.7 (Continued)

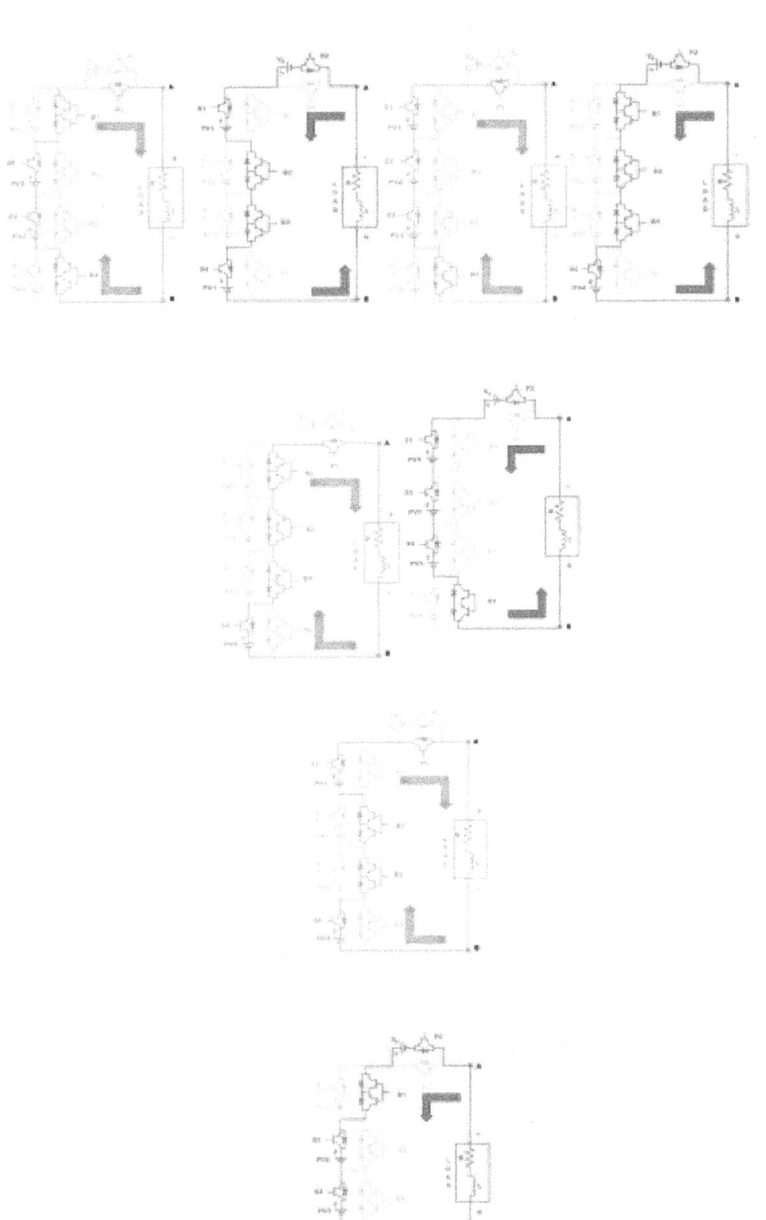

5V_{DC} -5V_{DC} 6V_{DC} -6V_{DC}
7V_{DC} -7V_{DC} 8V_{DC} -8V_{DC}

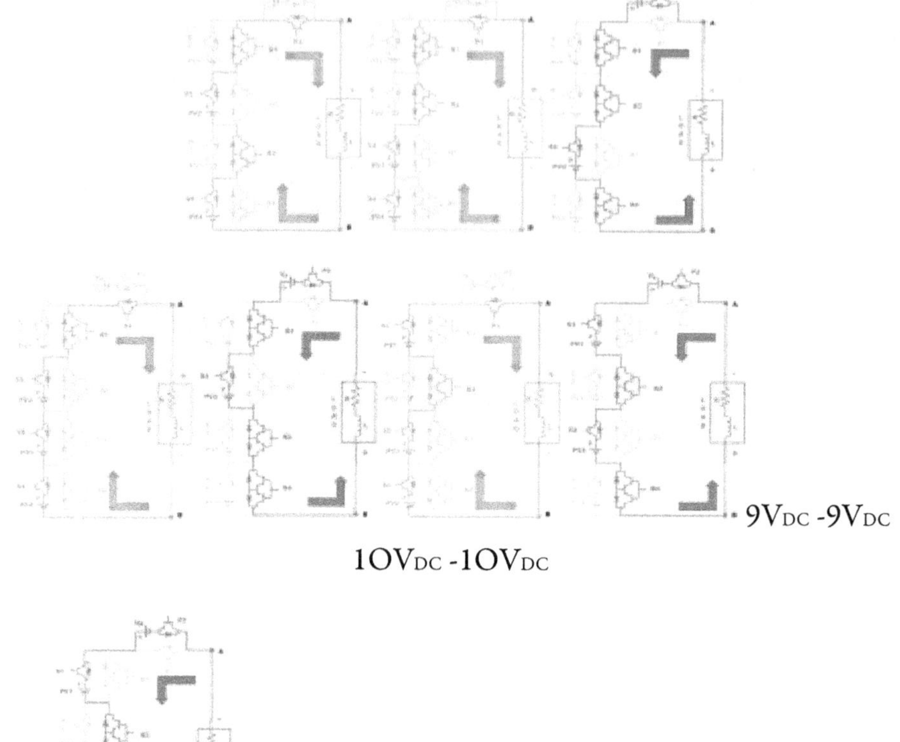

$9V_{DC}$ $-9V_{DC}$

$10V_{DC}$ $-10V_{DC}$

$11V_{DC}$ $-11V_{DC}$ $12V_{DC}$ $-12V_{DC}$

Figure 3.7 (Continued)

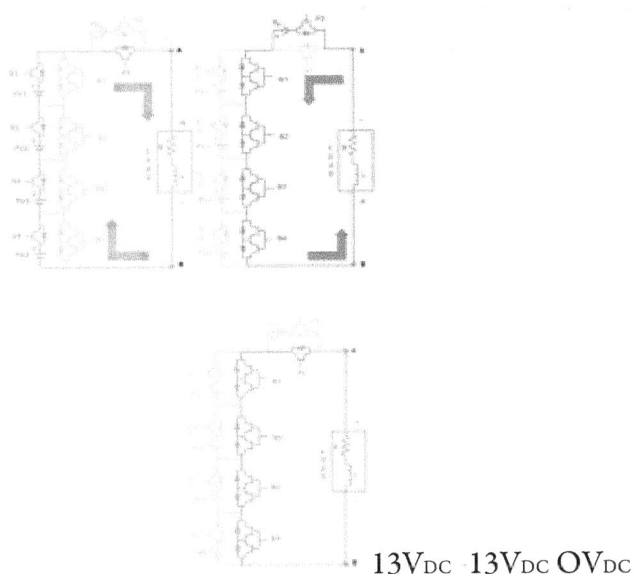

Figure 3.7 Modes of operation of Asymmetric 27 -level inverter

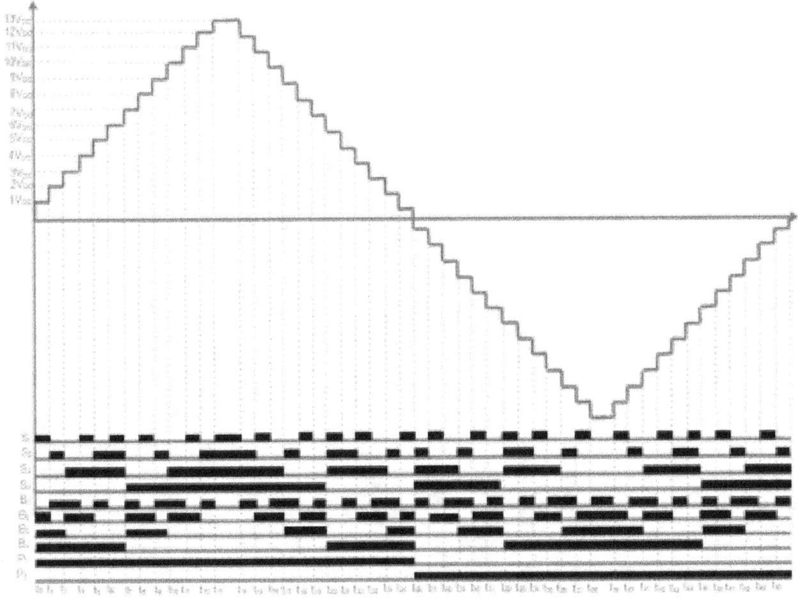

Figure 3.8 Switching operation of Asymmetric 27 -level inverter

Table 3.7 Current Path of 27 -Level MMI

Output voltage	Switches in conduction	Current path
+Vdc	S1, P1, B2, B3, B4	PV_{1+} S1 P1 Load B4 B3 B2 PV_{1-}
+2Vdc	S2, B1, P1, B3, B4	PV_{2+} S2 B1 P1 Load B4 B3 PV_{2-}
+3Vdc	S3, B2, B1, P1, B4	PV_{3+} S3 B2 B1 P1 Load B4 PV_{3-}
+4Vdc	S3, B2, S1, P1, B4	PV_{3+} S3 B2 PV_1 S1 P1 Load B4 PV_{3-}
+5Vdc	S3, S2, B1, P1, B4	PV_{3+} S3 PV_2 S2 B1 P1 Load B4 PV_{3-}
+6Vdc	S3, S2, S1, P1, B4	PV_{3+} S3 PV_2 S2 PV_1 S1 P1 Load B4 PV_{3-}
+7Vdc	S4, B3, B2, B1, P1	PV_{4+} S4 B3 B2 B1 P1 Load PV_{4-}
+8Vdc	S4, B3, B2, S1, P1	PV_{4+} S4 B3 B2 PV_1 S1 P1 Load PV_{4-}
+9Vdc	S4, B3, S2, B1, P1	PV_{4+} S4 B3 PV_2 S2 B1 P1 Load PV_{4-}
+10Vdc	S4, S3, B2, B1, P1	PV_{4+} S4 PV_3 S3 B2 B1 P1 Load PV_{4-}
+11Vdc	S4, S3, B2, S1, P1	PV_{4+} S4 PV_3 S3 B2 PV_1 S1 P1 Load PV_{4-}
+12Vdc	S4, S3, S2, B1, P1	PV_{4+} S4 PV_3 S3 PV_2 S2 B1 P1 Load PV_{4-}
+13Vdc	S4, S3, S2, S1, P1	PV_{4+} S4 PV_3 S3 PV_2 S2 PV_1 S1 P1 Load PV_{4-}

Table 3.7 (Continued)

Output voltage	Switches in conduction	Current path
0Vdc	B1, B2, B3, B4, P1	B1 P1 Load B4 B3 B2 B1
-Vdc	P2, B1, S2, S3, S4	V_{peak+} B1 S2 PV_2 S3 PV_3 S4 PV_4 Load P2 V_{peak-}
-2Vdc	P2, S1, B2, S3, S4	V_{peak+} S1 PV_1 B2 S3 PV_3 S4 PV_4 Load P2 V_{peak-}
-3Vdc	P2, B1, B2, S3, S4	V_{peak+} B1 B2 S3 PV_3 S4 PV_4 Load P2 V_{peak-}
-4Vdc	P2, B1, S2, B3, S4	V_{peak+} B1 S2 PV_2 B3 S4 PV_4 Load P2 V_{peak-}
-5Vdc	P2, S1, B2, B3, S4	V_{peak+} S1 PV_1 B2 B3 S4 PV_4 Load P2 V_{peak-}
-6Vdc	P2, B1, B2, B3, S4	V_{peak+} B1 B2 B3 S4 PV_4 Load P2 V_{peak-}
-7Vdc	P2, S1, S2, S3, B4	V_{peak+} S1 PV_1 S2 PV_2 S3 PV_3 B4 Load P2 V_{peak-}
-8Vdc	P2, B1, S2, S3, B4	V_{peak+} B1 S2 PV_2 S3 PV_3 B4 Load P2 V_{peak-}
-9Vdc	P2, S1, B2, S3, B4	V_{peak+} S1 PV_1 B2 S3 PV_3 B4 Load P2 V_{peak-}
-10Vdc	P2, B1, B2, S3, B4	V_{peak+} B1 B2 S3 PV_3 B4 Load P2 V_{peak-}
-11Vdc	P2, B1, S2, B3, B4	V_{peak+} B1 S2 PV_2 B3 B4 Load P2 V_{peak-}
-12Vdc	P2, S1, B2, B3, B4	V_{peak+} S1 PV_1 B2 B3 B4 Load P2 V_{peak-}
-13Vdc	P2, B1, B2, B3, B4	V_{peak+} B1 B2 B3 B4 Load P2 V_{peak-}

Table 3.8 Possible Levels generated by the AMMI

Configuration	Ratio of DC sources	5th DC source $\dfrac{N_L-1}{2}$	Output voltage levels N_L	Output voltage levels (cascaded) N_{Lc}	Equations on the basis of levels N_L
I	1:1:1:2	5	11	21	$N_{switch} = (N_L + 3)$ $N_{GatDri} = (N_L - 1)$ $(N_L - 1)$ $N_{DC} = 2$
II	1:1:2:2	6	13	25	$N_{switch} = (N_L + 1)$ $N_{GatDri} = (N_L - 3)$ $(N_L - 3)$ $N_{DC} = 2$
III	1:2:2:2	7	15	29	$N_{switch} = (N_L - 1)$ $N_{GatDri} = (N_L - 5)$ $(N_L - 5)$ $N_{DC} = 2$
IV	1:2:2:3	8	17	33	$N_{switch} = (N_L - 3)$ $N_{GatDri} = (N_L - 7)$ $(N_L - 7)$ $N_{DC} = 2$ $N_{switch} = (N_L - 5)$

V	1: 1: 2: 4	9	19	37	$N_{GatDri} = (N_L - 9)$ $(N_L - 9)$ $N_{DC} = 2$ $N_{switch} = (N_L - 7)$ $(N_L - 1)$
VI	1: 2: 3: 4	10	21	41	$N_{GatDri} = 2$ $(N_L - 1)$ $N_{DC} = 4$ $(N_L + 5)$ $N_{switch} = 2$
VII	1: 2: 3: 5	11	23	45	$(N_L - 3)$ $N =$ $_{GatDri}\ 2$ $(N_L - 3)$ $N_{DC} = 4$

_______Table 3.8 (Continued)

Configuration	Ratio of DC sources	5th DC source $\frac{N_L-1}{2}$	Output voltage levels N_L	Output voltage levels (cascaded) N_{Le}	Equations on the basis of levels N_L
VIII	1: 2: 3: 6	12	25	49	$N_{switch} = \dfrac{2}{(N_L + 3)}$ $N_{GatDri} = \dfrac{2}{(N_L - 5)}$ $N_{DC} = 4$
IX	1: 2: 3: 7	13	27	53	$N_{switch} = \dfrac{2}{(N_L + 1)}$ $N_{GatDri} = 3$ $(N_L + 3)$

$\underline{N_L\text{-}7}$

$N_{DC} = 4$

$(N_L - 1)$

$N_{switch} = 2$

$(N_L + 1)$

$N =$

GatDri 3

$\underline{N_L+1}$

$N_{DC} = 6$

$(N_L - 3)$

$N_{switch} = 2$

$(N_L - 1)$

$N =$

GatDri 3

$(N_L - 1)$

$N_{DC} = 6$

X

1:
2:
4: 14 29 57
7

XI

1:
2:
4: 15 31 61
8

———The configuration XI produces the maximum voltage (31-level) from 14 switches and 5 DC sources in all the above processes. In the suggested configuration the positive terminal of the DC supply has not been connected with the anode end of the diode. Occurrence of short-circuits in the suggested structure is not possible because of the usage of two-way switches and anti-parallel diodes. The proposed MMI configuration is a viable option for PV implementations. In general, the suggested MMIs performance in PV system is achieved by integrating several photovoltaic panels equipped with several boost converters Prabaharan & Palanisamy (2016).

COMPARISON OF THE PROPOSED 27-LEVEL AMMI WITH CONVENTIONAL 27-LEVEL INVERTER TOPOLOGIES

In order to validate the suggested structure on the basis of power electronic components, it should be compared with similar MLI configurations. Table 3.9 presents the comparison of the proposed 27-level AMMI with other 27-level asymmetric topologies. The comparison is done with respect to the voltage levels, semiconductor switches used, power diodes, drive circuits, and DC sources.

From Table 3.9, it is evident that for obtaining 27-level, the proposed reduced multilevel connections need only 14 IGBTs, 10 gate drivers, 14 power diodes, and 5 DC sources. For a conventional CHB inverter in the symmetric mode, 52 switches with 52 diodes,52 drive circuits, and 13 DC sources are needed for producing the 27- levels at the output.

Irusapparajan *et al.* (2019) proposed the trinary configuration of CHBMLI. It produces a 27-level of the output using 12 switches,12 power diodes,12 drive circuits, and 3 DC sources. When compared with the 27-level AMMI, the CHB trinary structure requires less DC sources and switches. But the limitation of more DC sources can be compensated by implementing a single input multiple output boost converter and in the CHB trinary structure the switches experience high voltage stress during conduction.

Table 3.9 Comparison of 27-Level Inverter Topologies

TOPOLOGY	N_L	N_{SWitch}	N_{Diode}	$N_{Gate\ Driver}$	N_{DC}	N_c	Negative levels
NPC	27	52	325	52	1	Nil	Two arms
FC	27	52	52	52	1	650	Two arms
CHB	27	52	52	52	13	Nil	H-bridge
Najafi & Yatim (2012)	27	30	30	30	13	Nil	H-bridge
Ebrahim et al. (2012)	27	30	30	13	17	Nil	H-bridge
Kangarlu et al. (2012)	27	28	28	18	13	Nil	Inherent
Babaei et al. (2012)	27	28	28	28	13	Nil	H-bridge
Gupta & Jain (2013)	27	28	28	28	13	Nil	Inherent
Ajami et al. (2014)	27	39	39	39	13	Nil	Inherent
Ajami et al. (2014)	27	29	29	29	13	Nil	H-bridge
Babaei et al. (2015)	27	26	26	26	13	Nil	H-bridge

| Alisha et al. (2015) | 27 28 | 28 | 16 | 13 | Nil | H-bridge |
| Prabaharan & palaniswamy (2016) | 27 14 | 14 | 12 | 5 | Nil | Inherent |

Table 3.9 (Continued)

TOPOLOGY	N_L	N_{SWitch}	N_{Diode}	$N_{Gate\ Driver}$	N_{DC}	N_c	Negative levels	Reduction in N_{SWitch}(%)
Babaei et al. (2016)	27	14	14	12	7	Nil	Inherent	0
Sathik et al. (2017)	27	18	18	18	7	12	H-bridge	22
Prabaharan et al. (2020)	27	45	45	35	11	Nil	H-bridge	69
Masoudinia et al. (2020)	27	16	16	12	5	Nil	Inherent	13
Proposed	27	14	14	10	5	Nil	Inherent	0

Let us consider the trinary CHBMLI structure with four DC sources (27V, 81V, 243V, 729V) as input. When the first voltage level is generated

the H-bridge switches will have a voltage stress of 1V ,but when the level

goes on increasing and when it reaches the fifth level, the H-bridge switches will experience 81V.This high voltage stress on the switches will damage the switches of H-bridge. To overcome this issue in CHBMLI, high rated

switches are implemented. The suggested MMI uses Common Emitter bilateral switches to decrease the losses. Moreover, the benefit of the suggested configuration is to produce positive and negative levels without H-bridge. The suggested configuration has a standardized distribution of stress on the switches. The sources feeding the CHBMLI are

V, 3V, 9V, 27V, 81V, 243V, 729V which are very unworkable to execute. In practice it needs to link numerous photovoltaic panels in sequence and includes a broad range of solar radiation absorption.

The differences in the above MLIs indicate that the suggested MLI, intensely decreases the quantity of switches and increases the outputs relative to all the above configurations. The reduction in the quantity of electronic switches, control components, implies that power losses and gross MLI costs have been minimized.

ASYMMETRIC 31-LEVEL MODULAR MULTILEVEL INVERTER TOPOLOGY

31-level inverter is obtained by taking the input DC source ratios as 1: 2: 4: 8: 15. The power switches are intelligently designed to get the possible voltage levels. Fifteen positive, fifteen negative and zero level combine to

form a 31-level. The positive fifteen-level is generated by triggering the switches S1 to S4, B1 to B4 randomly along with switch P1. To generate the negative fifteen-level in the suggested configuration, the switches S1 to S4, B1 to B4 are randomly switched along with switch P2. By implementing the zero-voltage triggering condition from Table 3.10 the zeroth level is produced. Table 3.11 presents the parameters of the suggested topology.

Table 3.10 Switching States of the Proposed 31-Level Inverter

S1	S2	S3	S4	B1	B2
✓	Ξ	Ξ	Ξ	Ξ	✓
Ξ	✓	Ξ	Ξ	✓	Ξ
✓	✓	Ξ	Ξ	Ξ	Ξ
Ξ	Ξ	✓	Ξ	✓	✓
✓	X	✓	Ξ	X	✓
X	✓	✓	Ξ	✓	Ξ

Table 3.10 (Continued)

	S1	S2	S3	S4	B1	B2
	✓	✓	✓	Ξ	X	Ξ
	X	X	Ξ	✓	✓	✓
	✓	X	Ξ	✓	X	✓
	X	✓	Ξ	✓	✓	X
	✓	✓	Ξ	✓	X	X
	X	X	✓	✓	✓	✓
	✓	X	✓	✓	X	✓
	X	✓	✓	✓	✓	X
	✓	✓	✓	✓	X	X
	Ξ	Ξ	Ξ	Ξ	✓	✓
	X	✓	✓	✓	✓	X
	✓	X	✓	✓	X	✓

X	X	✓	✓	✓	✓	Ξ
✓	✓	Ξ	✓	X	X	✓
X	✓	Ξ	✓	✓	X	✓
✓	X	Ξ	✓	X	✓	✓
X	X	Ξ	✓	✓	✓	✓
✓	✓	✓	Ξ	X	X	Ξ
X	✓	✓	Ξ	✓	X	Ξ
✓	X	✓	Ξ	X	✓	Ξ
X	X	Ξ	Ξ	✓	✓	Ξ
✓	✓	Ξ	Ξ	X	X	✓
X	✓	Ξ	Ξ	✓	X	✓
✓	X	Ξ	Ξ	X	✓	✓
X	X	Ξ	Ξ	✓	✓	Ξ

Table 3.11 Parameters of the 31-Level MMI

Parameters	Based on Number of Levels	Based on Number of Modules
Number of levels	N_L	$30n+1$
Number of switches	$\dfrac{N_L-3}{2}$	$13n+1$
Number of diodes	$\dfrac{N_L-3}{2}$	$13n+1$
Number of drive circuit	$\dfrac{N_L-1}{3}$	$10n$
Number of DC sources	$\dfrac{N_L-1}{6}$	$5n$
Total Standing Voltage (TSV)	$5\left[\dfrac{N_L-1}{2}\right]$	$75n$

The parameters presented in Table 3.11 is designed by considering the number of levels(N_L) and sectional units (n). Higher voltage levels can be generated by the suggested configuration when connected in cascaded form.

When two 31-level AMMI structures are connected in the cascaded manner a 61-level (30 positive level,30 negative level and zero level) is generated at the output. The modes of operation of 31-level AMMI is displayed in Figure 3.9 and the respective current path is shown in Table 3.12.

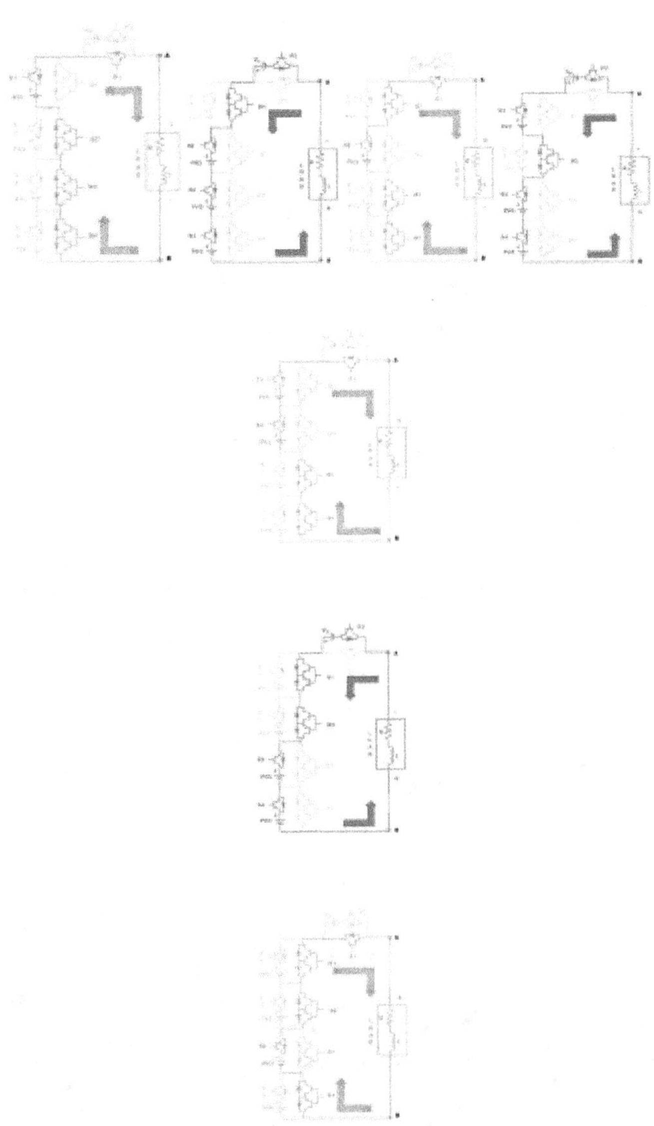

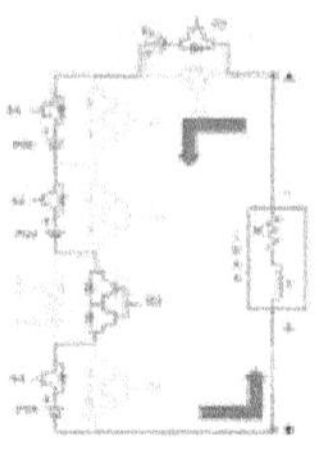

$$V_{DC} \;\text{-}V_{DC}\; 2V_{DC}\; \text{-}2V_{DC}$$
$$3V_{DC}\; \text{-}3V_{DC}\; 4V_{DC}\; \text{-}4V_{DC}$$

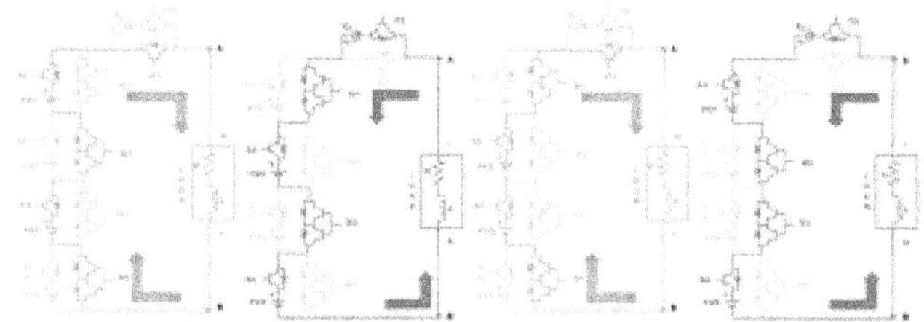

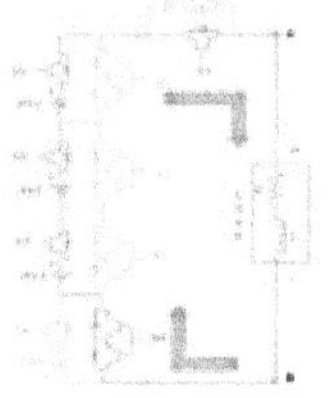

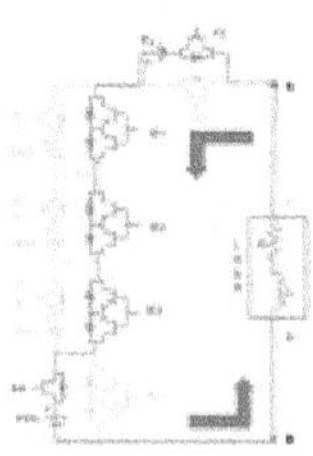

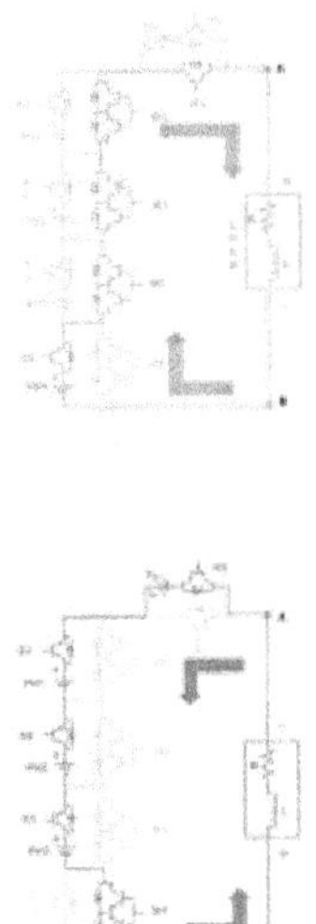

5V$_{DC}$ -5V$_{DC}$ 6V$_{DC}$ -6V$_{DC}$
7V$_{DC}$ -7V$_{DC}$ 8V$_{DC}$ -8V$_{DC}$

Figure 3.9 (Continued)

9V$_{DC}$ -9V$_{DC}$ 1OV$_{DC}$ -1OV$_{DC}$
11V$_{DC}$ -11V$_{DC}$ 12V$_{DC}$ -12V$_{DC}$

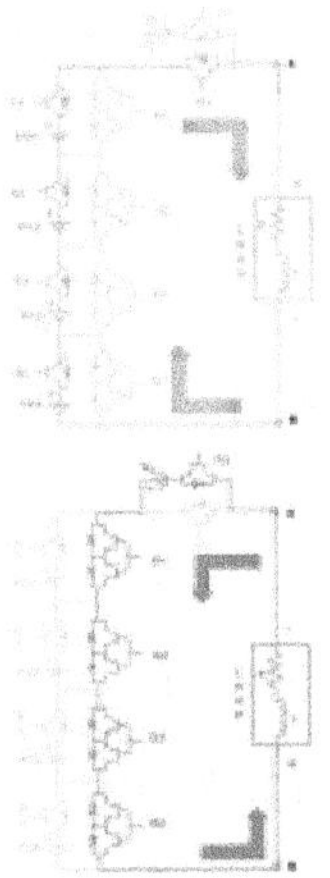

13V$_{DC}$ -13V$_{DC}$ 14V$_{DC}$ -14V$_{DC}$

15V$_{DC}$ -15V$_{DC}$ OV$_{DC}$

Figure 3.9 Modes of operation of Asymmetric 31-level inverter

Table 3.12 Current Path of 31-Level MMI

Output voltage	Switches in conduction	Current path
$+Vdc$	S1, P1, B2, B3, B4	PV_1+ S1 P1 Load B4 B3 B2 PV_1-
$+2Vdc$	S2, B1, P1, B3, B4	PV_2+ S2 B1 P1 Load B4 B3 PV_2-
$+3Vdc$	S1, S2, B3, P1, B4	PV_2+ S2 PV_1 S1 P1 Load B4 B3 PV_2-
$+4Vdc$	S3, B2, B1, P1, B4	PV_3+ S3 B2 B1 P1 Load B4 PV_3-
$+5Vdc$	S3, S2, B1, P1, B4	PV_3+ S3 B2 PV_1 S1 P1 Load B4 PV_3-
$+6Vdc$	S3, S2, B1, P1, B4	PV_3+ S3 PV_2 S2 B1 P1 Load B4 PV_3-
$+7Vdc$	S3, S2, S1, P1, B4	PV_3+ S3 PV_2 S2 PV_1 S1 P1 Load B4 PV_3-
$+8Vdc$	S4, B3, B2, B1, P1	PV_4+ S4 B3 B2 B1 P1 Load PV_4-
$+9Vdc$	S4, B3, B2, S1, P1	PV_4+ S4 B3 B2 PV_1 S1 P1 Load PV_4-
$+10Vdc$	S4, B3, S2, B1, P1	PV_4+ S4 B3 PV_2 S2 B1 P1 Load PV_4-
$+11Vdc$	S4, S3, B2, S1, P1	PV_4+ S4 PV_3 S3 B2 PV_1 S1 P1 Load PV_4-
$+12Vdc$	S4, S3, B2, B1, P1	PV_4+ S4 PV_3 S3 B2 B1 P1 Load PV_4-
$+13Vdc$	S4, S3, B2, S1, P1	PV_4+ S4 PV_3 S3 B2 PV_1 S1 P1 Load PV_4-
$+14Vdc$	S4, S3, S2, B1, P1	PV_4+ S4 PV_3 S3 PV_2 S2 B1 P1 Load PV_4-
$+15Vdc$	S4, S3, S2, S1, P1	PV_4+ S4 PV_3 S3 PV_2 S2 PV_1 S1 P1 Load PV_4-

Table 3.12 (Continued)

Output voltage	Switches in conduction	Current path
0Vdc	B1, B2, B3, B4, P1	B1 P1 Load B4 B3 B2 B1
-Vdc	P2, B1, S2, S3, S4	V_{peak+} B1 S2 PV_2 S3 PV_3 S4 PV_4 Load P2 V_{peak-}
-2Vdc	P2, S1, B2, S3, S4	V_{peak+} S1 PV_1 B2 S3 PV_3 S4 PV_4 Load P2 V_{peak-}
-3Vdc	P2, B1, B2, S3, S4	V_{peak+} B1 B2 S3 PV_3 S4 PV_4 Load P2 V_{peak-}
-4Vdc	P2, S1, S2, B3, S4	V_{peak+} S1 PV_1 S2 PV_2 B3 S4 PV_4 Load P2 V_{peak-}
-5Vdc	P2, B1, S2, B3, S4	V_{peak+} B1 S2 PV_2 B3 S4 PV_4 Load P2 V_{peak-}
-6Vdc	P2, S1, B2, B3, S4	V_{peak+} S1 PV_1 B2 B3 S4 PV_4 Load P2 V_{peak-}
-7Vdc	P2, B1, B2, B3, S4	V_{peak+} B1 B2 B3 S4 PV_4 Load P2 V_{peak-}
-8Vdc	P2, S1, S2, S3, B4	V_{peak+} S1 PV_1 S2 PV_2 S3 PV_3 B4 Load P2 V_{peak-}
-9Vdc	P2, B1, S2, S3, B4	V_{peak+} B1 S2 PV_2 S3 PV_3 B4 Load P2 V_{peak-}
-10Vdc	P2, S1, B2, S3, B4	V_{peak+} S1 PV_1 B2 S3 PV_3 B4 Load P2 V_{peak-}
-11Vdc	P2, B1, B2, S3, B4	V_{peak+} B1 B2 S3 PV_3 B4 Load P2 V_{peak-}
-12Vdc	P2, S1, S2, B3, B4	V_{peak+} S1 PV_1 S2 PV_2 B3 B4 Load P2 V_{peak-}
-13Vdc	P2, B1, S2, B3, B4	V_{peak+} B1 S2 PV_2 B3 B4 Load P2 V_{peak-}
-14Vdc	P2, S1, B2, B3, B4	V_{peak+} S1 PV_1 B2 B3 B4 Load

199

P2 V_{peak-}

V_{peak+} B1 B2 B3 B4 Load P2

-15Vdc   P2, B1, B2, B3, B4 V_{peak-}

Table 3.13 Comparison of the Proposed 31-Level Inverter with conventional 31-Level Inverter Structures

Structure	N_{Level}	N_{SWitch}	N_{Diode}	$N_{Gate\ Driver}$	N_{DC}
NPC	31	60	870	60	1
FC	31	60	60	60	1
CHB	31	60	60	60	15
Lee et al. (2002)	31	16	16	16	4
Su (2005)	31	34	34	34	15
Choi & Kang (2009)	31	32	32	32	15
Hinago & Koizumi (2010)	31	46	46	46	15
Babaei et al. (2012)	31	32	32	32	15
Menaka & Muralidharan (2017)	31	19	19	19	15

Gautam et al. (2018)	31	60	60	60	15
Dhanamjayulu et al. (2019)	31	15	15	12	4
Prabhu et al. (2019)	31	16	16	16	4
Mahato et al. (2019)	16	16	16	16	5
Proposed	31	14	14	10	5

POWER LOSSES CALCULATION

Loss calculation of a MMI is a dynamic task. MLI consist of semiconductor switches such as diodes, IGBTs, and MOSFETs which have specific power loss. Such losses are categorized into three different types namely triggering, blocking and conduction losses. The losses generated by an electronic switch during blocking a specific voltage is called blocking loss. Since the blocking loss is very less when equated with other two losses, it is not considered for loss calculation.

The complication in loss computation is due to the usage of specific current ratings in MMIs on semiconductor switches. Once the output is reached to a specific value, the switches corresponding to that particular arm will alter the current during the switching process. In addition, the complexity of the calculation of power loss is due to the frequency of switching. As the switch count is increased to fetch a greater amount of levels, the frequency of switching also varies across all switches. Therefore, loss computations also increase.

3.9.1 Conduction Losses

Conduction losses occur during the turn ON operation of an IGBT. The power distributed by an IGBT can be decided from its power during the ON state. The power dissipated by an IGBT can be obtained by the power consumed by the device during its turn on state.

In addition, the average dissipated power is achieved by manipulating the duty cycle with conduction loss. The proposed MMI contains two way switches in which its peak voltage is achieved by integrating the peak voltage of a one-way switch with two diodes. The conduction loss of a two-way switch is indicated as (ρcdn.IGBT) and is conveyed as:

$$p_{cdn(t)\ IGBT(B)} = (V_{ov} + 2V_{Di})I_k\sin wt + 2R_{Di}I_{k2}\sin^2 wt +$$

$$R_{ov}I_{k8+1}\sin^{8+1} wt \quad (3.2)$$

Where V_{ov} and V_{Di} are the voltages of IGBT and diode during conduction. the R_{ov} and R_{Di} are the resistances of IGBT and diode during conduction, I_k is the current passing through the switch and o is taken from

the datasheet of IGBT.

Likewise, the ON state loss of a unidirectional switch is expressed as

$$p_{cdn(t)\ IGBT(U)} = (V_{ov} + V_{Di})I_k\sin wt + R_{Di}I_{k2}\sin 2wt +$$

$$R_{ov}I_{k8+1}\sin^{8+1} wt \quad (3.3)$$

Over-all ON state loss of the AMMI is presented as

$$p_{cdn(t)} = .\ p_{cn(t)\ IGBT(B)} + v.\ p_{cdn(t)\ IGBT(U)} \quad (3.4)$$

Where both and v signify the total number of two-way and one-way switches during conduction. IGBT(u), IGBT(B) indicates the

unidirectional and bidirectional switches. From Figure 3.7 and Table 3.5 it is observed that 5 switches conduct for a full cycle.

Figure 3.8 describes the triggering states of 14 switches. The 3 two- way switches and 2 one-way switches operate from time period t_0 to t_1, t_1 to

t_2, t_2 to t_3 and t_6 to t_7. Similarly, 2 numbers of two-way and 3 numbers of

one-way switches operate from time period t_3 to t_4, t_4 to t_5, t_7 to t_8, t_8 to t_9 and t_{10} to t_{11}. A single unidirectional switch and 4 bidirectional switches functions from time period t_5 to t_6, t_{11} to t_{12} and t_{12} to t_{13}. Eventually, during the time period t13 to t14 five unidirectional switches are operated.

Thus, the average conduction loss of the suggested inverter is shown in Equation (3.5) for the quarter half-cycle.

$$\overline{p}$$

p 4 t

cdn(t) = T [∫to 3pcdn(t)IGBT(B) + 2pcdn(t)IGBT(U) dt +

t2 3p

+ 2p

dt +

t3 3p

+ 2p

dt +
∫t cdn(t)IGBT(B)

cdn(t)IGBT(U)

∫t2 cdn(t)IGBT(B)

cdn(t)IGBT(U)

t7 3p

+ 2p

dt +

t4 2p

+ 3p

dt +
∫t6 cdn(t)IGBT(B)

cdn(t)IGBT(U)

J_{t3} cdn(t)IGBT(B)

cdn(t)IGBT(U)

$_{ts} 2p$

$+ 3p$

$dt +$

$_{t8} 2p$

$+ 3p$

$dt +$

J_{t4}

cdn(t)IGBT(B)

cdn(t)IGBT(U)

J_{t7}

cdn(t)IGBT(B)

cdn(t)IGBT(U)

$_{t9} 2p$

$+ 3p$

$dt +$

$_t 2p$

$+ 3p$

$dt +$

J_{t8}

cdn(t)IGBT(B)

cdn(t)IGBT(U)

J_t o

cdn(t)IGBT(B)

cdn(t)IGBT(U)

$$+ 4p \qquad dt \quad + \quad^{\tau\ 2} \qquad + 4p \qquad \text{GET} \quad (u$$
$$1p \qquad\qquad\qquad\qquad\qquad) \ dt \quad +$$
$$+ \qquad dt \quad +$$
$$4p \qquad\qquad\qquad\qquad (3.5$$
$$dt \qquad\qquad\qquad\qquad)$$
$$t6\ 1p$$

J_{ts} cdn(t)IGBT(B)

cdn(t)IGBT(U)

J_t

cdn(t)IGBT(B)

cdn(t)I
  $t3\ 1p$

J_{t2}

cdn(t)IGBT(B)

cdn(t)IGBT(U)
  $J_t 4\ 5p$

$t3$

3.9.2

cdn(t)IGBT(U)

The loss is correlated with the switch as well as the anti-parallel diode while considering an IGBT. The switching frequency is quite directly proportionate to the triggering loss. The triggering losses are therefore mainly known to measure the overall inverter loss. Almost 5 significant features have an effect on switch behaviour such as (1) the blocking voltage of the switch,

1. the gate resistance, (3) the stray inductance, (4) the current flowing through the switch and (5) the temperature present in the junction of a semiconductor. The losses produced from the semiconductor switches for different applications is commonly considered to be a primary disadvantage for multilevel inverters. Moreover, these losses increase the operating cost and lower inverter performance.

There are 6 one-way switches and 4 two-way switches in the proposed topology. Triggering losses are mainly related with the triggering frequency of the one-way (f_U) and two-way switches (f_B). The triggering frequencies of the electronic switches are achieved from Figure 3.8.

Furthermore, from Figure 3.8 it is detected that the switches P_1 and P_2 work at the fundamental frequency (f_f) which is 50 Hz.

Switching loss of the suggested inverter is given in Equation 3.6.

$$p_{sw} = p_{sw(U)} + p_{sw(B)}$$

244
1 1 1

$$\_= L\,[$$
$$6$$
$$i=1$$

$$V_U I_L T f_f] + L\,[$$

$$\_6$$
$$i=1$$

$$V_U I_L T f_U] + L\,[$$

$$\_6$$
$$i=1$$

$$V_B I_L T f_B] \quad (3.6)$$

Where, V_U, V_B are the standing voltages of one-way and two-way switches, I_L is the Root Mean Square of the load current, $T = t_{on} + t_{off}$. Equation 3.6 can be modified further by taking $\frac{T * I}{L} =$. Therefore, Equation

$$6$$

3.6 becomes

244
$$p_{sw} = L\,V_U f_f + L\,V_U f_U + L\,V_B f_B \quad (3.7)$$

$$i=1$$

$i=1$

$i=1$

The switching losses are ignored for the switches operating under fundamental frequency. In addition, the common emitter structured two-way switch produces very low switching loss. Consequently, the switching loss in AMMI are relatively very small when equated with other asymmetric configurations.

3.9.3 Loss Calculation of the Proposed 27-level MMI

_The experimental output voltage and current of the proposed 27-level MMI is 312V and 7A. This is an average value. To convert the average value to RMS value, the obtained average value is divided by $\sqrt{2}$

.Therefore the RMS output value of the voltage and current of MMI is 220V

and 4.94A. The output power of the inverter is calculated using Equation 3.8.

$$P_o = V_{rms} * I_{rms} \quad (3.8)$$

$$= 220 * 4.94$$

$$= 1086.8 \text{ W}$$

The FGA25N120 IGBT (14nos) are used for designing the proposed MMI. The parameters required for calculating the conduction and switching loss are obtained from the data sheet. The onstage voltage drop is

found to be 0.6 V which is determined from the characteristic plot found in

the datasheet. Further the resistance of IGBT is 0.4 .0 ,turn-on delay time is 50 ns , rise time is 60 ns , turn-off delay time is 200 ns , and fall time is 154 ns .There will be 53 steps in the output waveform for one complete

cycle of the suggested 27-level MMI.

The conduction loss is given as
$$p_{cdn(t)} = (0.6 * 0.4(4.94)) * 4.94 * 14 = 81.99\text{W}$$

The switching loss is given as

$$p_{sw} = E_{on} + E_{off} \quad (3.9)$$

$$E_{on} = 220 * 4.94((50 + 60) * 10^{-9}) * 53 * 14 = 0.088W$$

E_{off} = 220 * 4.94((200 + 154) * 10^{-9}) * 53 * 14 = 0.2854W p_{sw} = 0.3734W

Total loss p = p_{cdn} + p_{sw} = 81.99W + 0.3734W = 82.3634W

Efficiency of the inverter is given in Equation 3.10.

$$\_\_\_\_\_1] = P_o$$

$$P_o + p$$

$$* 100 \ (3.10)$$

$$1086.8$$

$$\_\_\_\_\_\_\_ =$$

$$1086.8 + 82.3634$$

$$1086.8$$

$$\_\_\_\_\_ =$$

$$1169.1634$$

= 93%

The total loss generated by 27-level AMMI is 0.082pu. The topologies proposed by Ajami *et al.* (2014a), Babaie *et al.* (2015a), Sathik *et al.* (2017) generate a total loss of 2.29pu, 2.46pu, 1.47pu.

Figure 3.10 graphically depicts the comparison of losses of 27-level AMMI with other 27-level asymmetric structures.

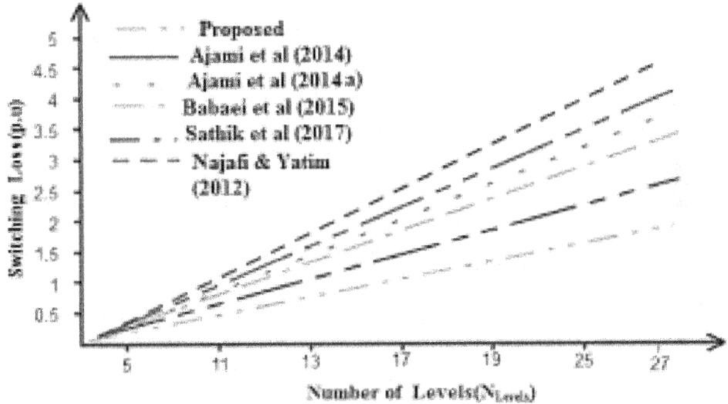

(a)

Figure 3.10 (Continued)

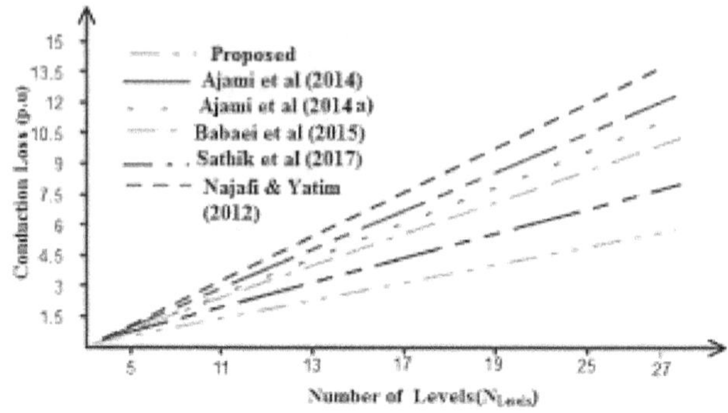

(b)

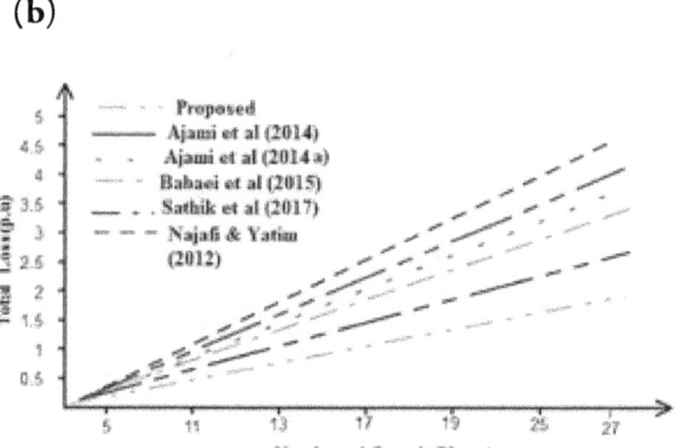

(c)

Figure 3.10 Comparison of (a) Switching loss (b) Conduction loss (c)Total loss for 27-level AMMI with Conventional 27-Level Configurations

3.10 SELECTIVE HARMONIC PULSE WIDTH MODULATION FOR THE PROPOSED MMI

Selective Harmonic Elimination (SHE) is a preprogramed PWM approach in which the triggering angles are computed for various modulation indices and are saved in a lookup table for the modulation of the MMI.SHE method is obtained from Fourier transform of a square wave which has various triggering states in a full cycle. By implementing this method, it is expected to determine that, for an inverter output voltage with 'S' triggering angles per quarter cycle, 'S' degrees of freedom are available for computation in the waveform. These degrees of freedom can be computed to confirm that a fundamental voltage component without (S-1) harmonics can be obtained.

These degrees of freedom are utilized by creating a set of equations to determine when the triggering of switches should occur in the AC waveform. But the equations obtained are nonlinear and transcendental in nature and so must be calculated using mathematical methods. Since there are drawbacks in calculating the switching angles using mathematical methods the optimization techniques such as Bee Colony Optimization (BCO), Particle Swarm Optimization (PSO) and Genetic Algorithm (GA) are used to calculate the triggering angles for the symmetric 9-level and asymmetric 27-level,31 level MMI.

For a N-level inverter 'S_{ang}' triggering angles are obtained to reduce 'S_{ang}-1' harmonics. Therefore, for the 9-level MMI the triggering angles are obtained using the Equation 1.1.3 given in chapter 1.

$$S_{ang} = \frac{9-1}{2} = 4$$

Therefore, for the 9-level MMI four triggering angles (θ_1, θ_2, θ_3, θ_4) are obtained by solving four nonlinear equations to reduce the 3_{rd},5_{th}

and 7_{th} order of harmonics. Equations 3.11-3.14 shows the nonlinear equations of the proposed 9-level MMI.

The nonlinear equation of the fundamental component is written as:

$$V = 4V_{pv} \cos(\theta) + \cos(\theta) + \cos(\theta) + \cos(\theta) \quad (3.11)$$

$$\overline{{}_1 \text{TI} {}_1}$$

234

220

The nonlinear equations for the odd harmonic component are written as:

$$V = 4V_{pv} \cos(3\theta_1) + \cos(3\theta_2) + \cos(3\theta_3) + \cos(3\theta_4) \quad (3.12)$$

$$\frac{}{3} 3TI_{1\,2\,3\,4}$$

$$V = 4V_{pv} \cos(5\theta_1) + \cos(5\theta_2) + \cos(5\theta_3) + \cos(5\theta_4) \quad (3.13)$$

$$\frac{}{5} 5TI_{1\,2\,3\,4}$$

$$V = 4V_{pv} \cos(7\theta_1) + \cos(7\theta_2) + \cos(7\theta_3) + \cos(7\theta_4) \quad (3.14)$$

$$\frac{}{7} 7TI_{1\,2\,3\,4}$$

3.10.1 Determination of Switching Angles using Newton Raphson Method

The Newton-Raphson (NR) approach is implemented to calculate the non-linear equations by means of a successive approximation technique. The estimation method contains the Equations (3.11) to (3.14).

Step 1: The switching angle matrix is defined in Equation (3.15) for the calculation of uncertain switching angles 1 to 4.

$$\theta^i = [\theta^1, \theta^2, \theta^3, \theta^4]_T \quad (3.15)$$

Step 2: Construction of nonlinear system matrix as given in Equations (3.16) and (3.17).

$$\cos(\theta^i) + \cos(\theta^i) + \cos(\theta^i) + \cos(\theta^i)$$

$$r_{1\,2\,3\,4}1$$

$$\cos(3\theta^i) + \cos(3\theta^i) + \cos(3\theta^i) + \cos(3\theta^i)$$

$$F_i = {}_{1\,2}$$

$$_{3\,4}(3.16)$$

$$\cos(5\theta_1^i) + \cos(5\theta_2^i) + \cos(5\theta_3^i) + \cos(5\theta_4^i)$$

$$\lceil \cos(7\theta_1^i) + \cos(7\theta_2^i) + \cos(7\theta_3^i) + \cos(7\theta_4^i) \rfloor$$

$$-\sin(\theta_1^i) - \sin(\theta_2^i) - \sin(\theta_3^i) - \sin(\theta_4^i)$$

$$\mathbf{r}$$

$$F \quad -\sin(3\theta_1^i) - \sin(3\theta_2^i) - \sin(3\theta_3^i) - \sin(3\theta_4^i)$$

$$\underline{\quad}[\;] = \qquad (3.17)$$

$$\theta \quad -\sin(5\theta_1^i) - \sin(5\theta_2^i) - \sin(5\theta_3^i) - \sin(5\theta_4^i)$$

$$\lfloor -\sin(7\theta_1^i) - \sin(7\theta_2^i) - \sin(7\theta_3^i) - \sin(7\theta_4^i) \rfloor$$

Step 3: Construction of the harmonic amplitude matrix as per the expression given in Equation (3.18).

$$M_i\,TI^{\,T}$$
$$\underline{\quad\quad}T = [\,4\;0^*0^*\;0\,]$$

$$(3.18)$$

Equation (3.11) and (3.18) can be modified as per Equation (3.19).

$$F(\text{\ding{117}}) = T \quad (3.19)$$

Using the matrices given in Equations (3.15) - (3.19) and NR method, using MATLAB software the described statement of algorithm is to be implemented into the program.

Step 1: Predict for initial values as per Equation (3.20). The method of

obtaining the initial predicted value by equal area criterion is given by Wang & Ahmadi (2010).

$$\text{\ding{117}}^0 = \text{\ding{117}}^0,\ \text{\ding{117}}^0,\ \text{\ding{117}}^0,\ \text{\ding{117}}^0 \quad (3.20)$$

1 2 3 4

Step 2: This step includes Equations (3.21) to (3.25).

$$F(\text{\ding{117}}_0) = F^0 \quad (3.21)$$

Linearizing the Equation (3.19) at $f\,]^0$ get,

and

$$F^0$$

$$F^0 + [\ ]$$

$$d\Phi^0 = T \quad (3.22)$$

$$d\Phi^0 = d\Phi^0_1, d\Phi^0_2, d\Phi^0_3, d\Phi^0_4 \quad (3.23)$$

Solving Equation (3.23) using the inverse of the matrix which is given in the Equation (3.24)

$$d\Phi^0 = \text{INV}\,[$$

$$\frac{F^0}{}\,]$$

$$T\text{-}F^0 \quad (3.24)$$

$$\Phi$$

Step 3: From Equation (3.25) the initial values are Updated.

$$\Phi^{i+1} = \Phi^i + df\,]^i \quad (3.25)$$

The steps (2) (3) are reiterated until $df\,]^i$ is satisfied to the degree of accuracy and the condition to be satisfied is Φ , Φ , Φ , Φ < TI.

12342

Similarly, for the remaining 11 configurations various nonlinear equations are solved to obtain the triggering angles for switching the MMI. These triggering angles are stored in a lookup table and executed accordingly.

3.11 SIMULATION RESULTS

The operation of the suggested MMI structure is validated by simulation using MATLAB / Simulink and experimental setup for both symmetric and asymmetrical conditions. The proposed asymmetric MMI produces a better level of output voltage than the symmetric MMI, which is verified in both simulation and experimental results.

3.11.1 9-Level MMI

SHE method is used to produce the pulses for the MMI. In order to examine and test the authority of proposed topology for a 9-level inverter, it is simulated by using MATLAB/ Simulink (2017a) simulation tool. In version 2017a the PV array block is readily available in the renewable energy library. Inside the PV array block the module parameters can be set by implementing the user defined option. Table 3.14 shows the dynamics used for the simulation of the MMI. The parameters presented in Table 3.14 are fed to produce12V ,7A output from each PV panel. For the symmetric MMI each PV source consists of four PV panels connected in series.

Table 3.14 Simulation Parameters

Parameters		Values
Number of Unidirectional Switches		6
Number of Bidirectional Switches		4
PV Array	Cells per module	24
	Voltage at Maximum Power Point V_{mp} (V)	12V
	Current at Maximum Power Point I_{mp} (A)	7A
	Open Circuit Voltage V_{oc} (V)	14.5V
	Short-Circuit Current I_{sc} (A)	7.3A
Ratio of DC sources		1: 1: 1: 1, 1: 1: 2: 2, 1: 2: 2: 2, 1: 1: 3: 3, 1: 1: 2: 4, 1: 2: 3: 4,1: 2: 3: 5, 1: 2: 3: 6, 1: 2: 3: 7, 1: 2: 4: 7, 1: 2: 4:8
Value of each DC source		For Symmetric V_{PV} = 50V For Asymmetric V_{PV} = 24V
Number of output voltage level		9-Level,11-Level, 13-Level, 15-Level, 17-Level,19-Level,21-Level,25-Level ,27-Level,29-Level and 31-Level.
Load (R & RL)		R = 100 .0 and R = 100.0 , L = 10mH

Therefore 48V,7A is obtained as the output for the PV source. This output is fed as the input to the DC-DC boost converter to give a constant 50V DC voltage to the MMI.

The fifth DC source is obtained by connecting the four PV panels in series. The peak voltage of the 9-level symmetric MMI is 200V $(4 * 50 =$

$200V)$ with a load value of R = 100.0 and L = 10mH. The corresponding

simulated 9-level MMI output voltage with its harmonic spectrum is presented in Figure 3.11. The corresponding voltage THDs is 7.54 %.

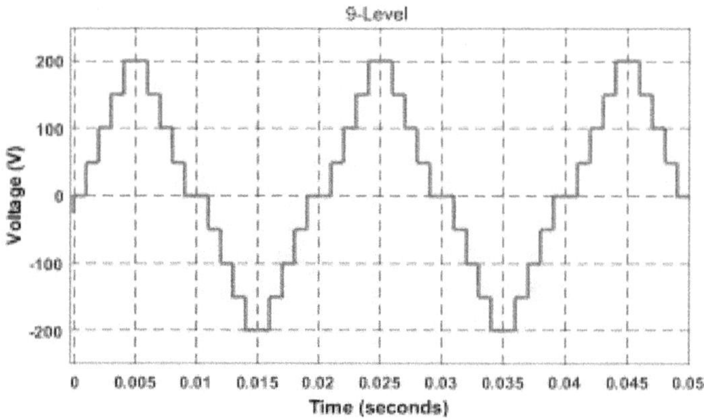

(a)

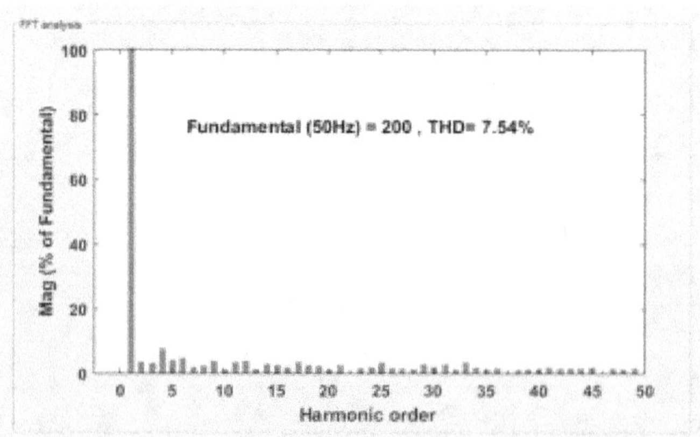

(b)

Figure 3.11 (a) 9-Level output voltage (b)FFT analysis of 9-Level Inverter

3.11.2 11-Level MMI

In Figure 3.2a, if the inputs are in 1: 1: 1: 2 ratio a 11-level output is

generated. Table 3.15 specifies the triggering sequence of the suggested 11- level AMMI. For the asymmetric MMI, the PV sources are different in nature so based on the PV inputs the PV panels are connected in series. For

simulation purpose the PV sources are taken as $V_{PV} = 24V$, $V_{PV2} =$

24V, $V_{PV3} = 24V$, $V_{PV4} = 48V$ (24: 24: 24: 48). The 24 V and 48 V is

obtained by connecting two and four 12V PV panels in series.

Table 3.15 Triggering Sequence for 11-Level AMMI

S1	S2	S3	S4	B1	B2	B3	B4	P1	P2	Output voltage (Vo)
✓	□	□	□	□	✓	✓	✓	✓	□	1 Vdc
✓	✓	□	□	□	□	✓	✓	✓	□	2 Vdc
✓	✓	✓	□	□	□	X	✓	✓	□	3 Vdc
✓	✓	□	✓	□	□	✓	X	✓	□	4 Vdc
✓	✓	✓	✓	□	□	□	□	✓	□	5 Vdc
□	□	□	□	✓	✓	✓	✓	✓	□	0 Vdc
□	✓	✓	✓	✓	□	□	□	X	✓	-1 Vdc
□	□	✓	✓	✓	✓	□	□	X	✓	-2 Vdc
□	□	□	✓	✓	✓	✓	□	X	✓	-3 Vdc
✓	□	□	□	□	✓	✓	✓	X	✓	-4 Vdc
□	□	□	□	✓	✓	✓	✓	□	✓	-5 Vdc

Figure 3.12 presents the simulation output and THD of the proposed 11-level AMMI. From Figure 3.12 it is observed that the peak inverter voltage is $V_{peak} = 120V$ and the output voltage THD obtained is 6.87%.

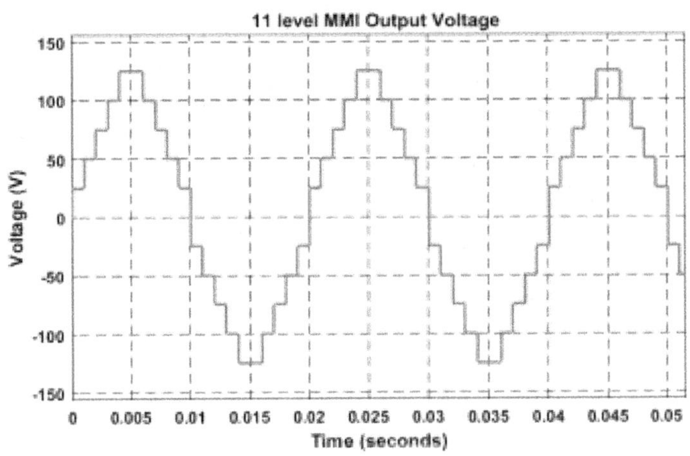

(a)

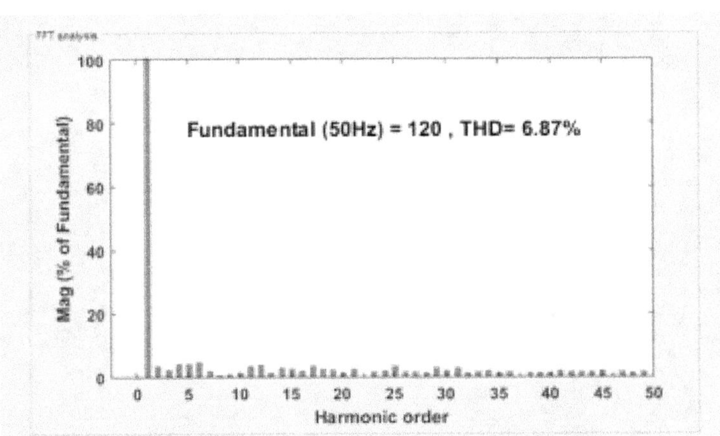

(b)

Figure 3.12 (a) 11-Level output voltage (b)FFT analysis of 11-Level Inverter

3.11.3 13-Level MMI

From the generalized Figure in 3.2a, if the inputs are in the ratio of 1: 1: 2: 2 a 13-level output is produced from the proposed structure. Table 3.16 shows the triggering sequence of the suggested 13-level AMMI. For simulation purpose the PV sources are taken as $V_{PV} = 24V$, $V_{PV2} =$

24V, $V_{PV3} = 48V$, $V_{V4} = 48V$ (24: 24: 48: 48). Figure 3.14 presents the

simulation output of the suggested 13-level AMMI. Figure 3.14 indicates the peak output voltage and THD as $V_{peak} = 144V$ and 6.12%.

Table 3.16 Triggering Sequence for 13-level AMMI

S1	S2	S3	S4	B1	B2	B3	B4	P1	P2	Output voltage (V
✓	Ξ	Ξ	Ξ	Ξ	✓	✓	✓	✓	Ξ	1 Vdc
X	X	✓	Ξ	✓	✓	X	✓	✓	Ξ	2Vdc
X	✓	✓	Ξ	✓	Ξ	X	✓	✓	Ξ	3 Vdc
X	X	✓	✓	✓	✓	X	X	✓	Ξ	4 Vdc
Ξ	✓	✓	✓	✓	Ξ	Ξ	Ξ	✓	Ξ	5 Vdc
✓	✓	✓	✓	Ξ	Ξ	Ξ	Ξ	✓	Ξ	6 Vdc
Ξ	Ξ	Ξ	Ξ	✓	✓	✓	✓	✓	Ξ	0 Vdc
Ξ	✓	✓	✓	✓	Ξ	Ξ	Ξ	X	✓	-1 Vdc
Ξ	Ξ	✓	✓	✓	✓	X	X	X	✓	-2Vdc
X	✓	✓	Ξ	✓	X	X	✓	X	✓	-3 Vdc
Ξ	Ξ	✓	Ξ	✓	✓	X	✓	X	✓	-4 Vdc
X	✓	X	X	✓	X	✓	✓	X	✓	-5 Vdc
Ξ	Ξ	Ξ	Ξ	✓	✓	✓	✓	X	✓	-6 Vdc

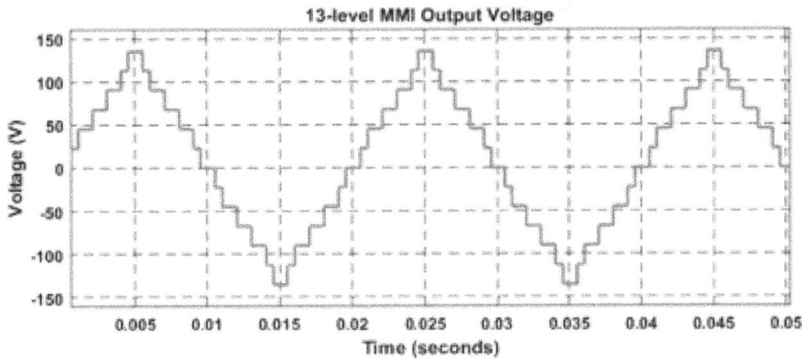

(a)

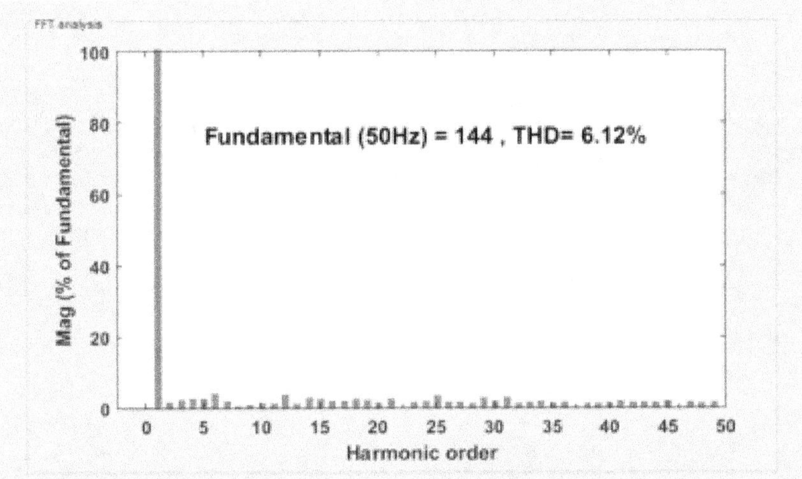

(b)

Figure 3.13 (a) 13-Level output voltage (b)FFT analysis of 13-Level Inverter

3.11.4 15-Level MMI

In Figure 3.2a if the inputs are in the ratio of 1: 2: 2: 2 a 15-level output is produced from the proposed structure. For simulation purpose the DC sources are taken as $V_{pv} = 24V$, $V_{pv2} = 48V$, $V_{pv3} = 48V$, $V_{pv4} = 48V$

(24: 48: 48: 48). Table 3.17 displays the switching sequence of the suggested

15-level MMI. Figure 3.15 presents the simulation output of the suggested 15- level AMMI with its harmonic spectrum. The output THD obtained is 5.90%.

Table 3.17 Triggering Sequence for 15-Level AMMI

S1	S2	S3	S4	B1	B2	B3
✓	☒	☒	☒	☒	✓	✓
☒	✓	☒	☒	✓	☒	✓
✓	✓	☒	☒	☒	☒	✓
☒	✓	✓	☒	✓	☒	☒
✓	✓	✓	☒	☒	☒	☒
☒	✓	✓	✓	✓	☒	☒
✓	✓	✓	✓	☒	☒	☒
☒	☒	☒	☒	✓	✓	✓
☒	✓	✓	✓	✓	☒	☒
✓	✓	✓	☒	☒	☒	☒
☒	✓	✓	☒	✓	☒	☒
✓	✓	☒	☒	☒	☒	✓
☒	✓	☒	☒	✓	☒	✓
✓	☒	☒	☒	☒	✓	✓
☒	☒	☒	☒	✓	✓	✓

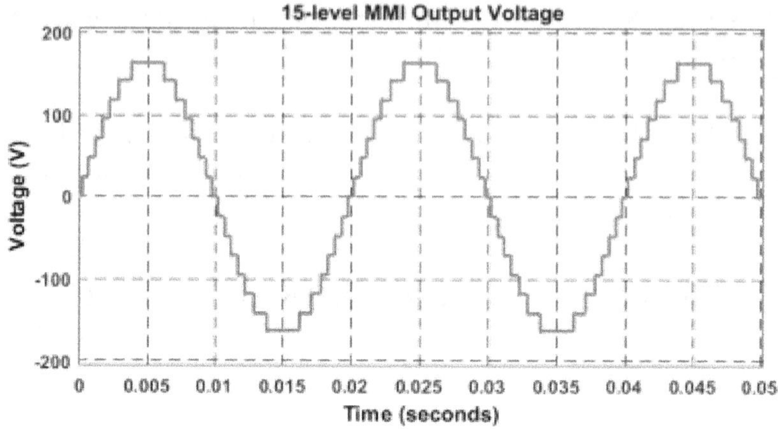

(a)

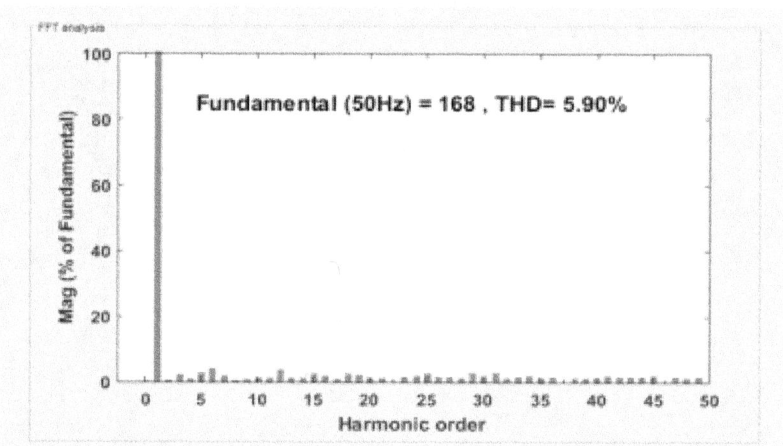

(b)

Figure 3.14 (a) 15-Level output voltage (b)FFT analysis of 15-Level Inverter

3.11.5 17-Level MMI

In Figure 3.2a if the inputs are in 1: 1: 3: 3 ratio a 17-level output is produced from the suggested structure. For simulation purpose the DC sources are taken as $V_{pv} = 24V$, $V_{pv2} = 24V$, $V_{pv3} = 72V$, $V_{pv4} = 72V$

(24: 24: 72: 72). Table 3.18 illustrates the triggering sequence of the

suggested 17-level MMI. Figure 3.15 shows the simulation output of the suggested 17-level AMMI. Figure 3.15 clearly indicates the peak output

voltage and THD as $V_{peak} = 192V$ and 5.79%.

Table 3.18 Triggering sequence for 17-Level AMMI

S1	S2	S3	S4	B1	B2	B3	B4	P1	P2	OUTPUT
✓	Ξ	Ξ	Ξ	Ξ	✓	✓	✓	✓	Ξ	1 Vdc
✓	✓	Ξ	Ξ	Ξ	Ξ	✓	✓	✓	Ξ	2Vdc
Ξ	Ξ	✓	Ξ	✓	✓	Ξ	✓	✓	Ξ	3 Vdc
✓	Ξ	✓	Ξ	Ξ	✓	Ξ	✓	✓	Ξ	4 Vdc
✓	✓	✓	Ξ	Ξ	Ξ	Ξ	✓	✓	Ξ	5 Vdc
Ξ	Ξ	✓	✓	✓	✓	Ξ	Ξ	✓	Ξ	6 Vdc
✓	Ξ	✓	✓	Ξ	✓	Ξ	Ξ	✓	Ξ	7 Vdc
✓	✓	✓	✓	Ξ	Ξ	Ξ	Ξ	✓	Ξ	8Vdc
Ξ	Ξ	Ξ	Ξ	✓	✓	✓	✓	✓	Ξ	0 Vdc
✓	Ξ	✓	✓	Ξ	✓	Ξ	Ξ	Ξ	✓	-1 Vdc
Ξ	Ξ	✓	✓	✓	✓	Ξ	Ξ	Ξ	✓	-2Vdc
✓	1	✓	Ξ	Ξ	Ξ	Ξ	✓	Ξ	✓	-3 Vdc
✓	Ξ	✓	Ξ	Ξ	✓	Ξ	✓	Ξ	✓	-4 Vdc
Ξ	Ξ	✓	Ξ	✓	✓	Ξ	Ξ	Ξ	✓	-5 Vdc
✓	✓	Ξ	Ξ	Ξ	Ξ	✓	✓	Ξ	✓	-6 Vdc
✓	Ξ	Ξ	Ξ	Ξ	✓	✓	✓	Ξ	✓	-7 Vdc
Ξ	Ξ	Ξ	Ξ	✓	✓	✓	✓	Ξ	✓	-8Vdc

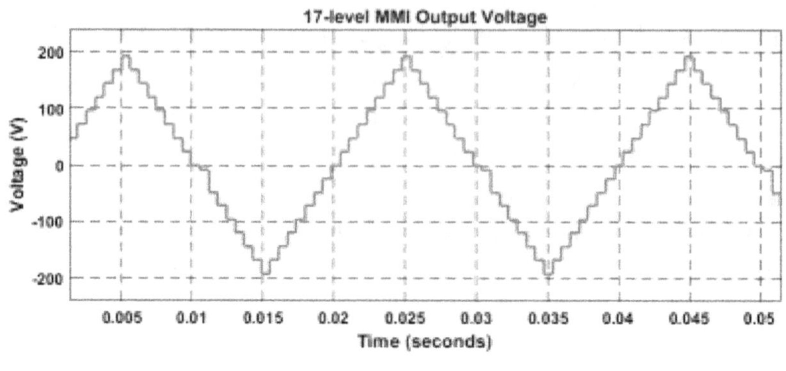

(a)

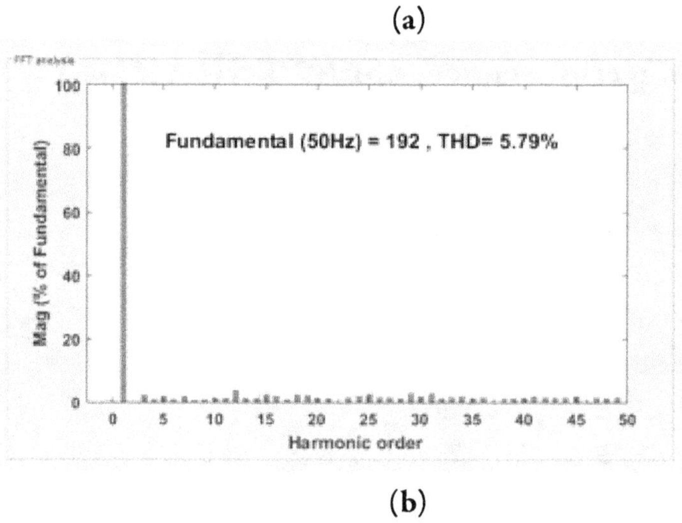

(b)

Figure 3.15 (a) 17-Level output voltage (b)FFT analysis of 17-Level Inverter

3.11.6 19-Level MMI

In Figure 3.2a if the inputs are in the ratio of 1: 2: 2: 4 a 19-level output is produced from the proposed structure. For simulation purpose the DC sources are taken as $V_{pv} = 24V$, $V_{pv2} = 48V$, $V_{pv3} = 48V$, $V_{pv4} = 96V$

(24: 48: 48: 96). Table 3.19 displays the triggering sequence of the suggested

19-level AMMI. Figure 3.16 illustrates the simulation output of the suggested

19-level AMMI. Figure 3.16 indicates the peak output voltage and THD as

V_{peak} = 216V and 5.32%.

Table 3.19 Triggering sequence for 19-Level AMMI

S1	S2	S3	S4	B1	B2	B3	B4	P1	P2	OUTPUT
✓	Ξ	Ξ	Ξ	Ξ	✓	✓	✓	✓	Ξ	1 Vdc
Ξ	✓	Ξ	Ξ	✓	Ξ	✓	✓	✓	Ξ	2 Vdc
✓	✓	Ξ	Ξ	✓	✓	Ξ	✓	✓	Ξ	3 Vdc
Ξ	Ξ	Ξ	✓	✓	✓	✓	Ξ	✓	Ξ	4 Vdc
✓	Ξ	Ξ	✓	Ξ	✓	✓	Ξ	✓	Ξ	5 Vdc
Ξ	✓	Ξ	✓	✓	Ξ	✓	Ξ	✓	Ξ	6 Vdc
✓	Ξ	✓	✓	Ξ	✓	Ξ	Ξ	✓	Ξ	7 Vdc
Ξ	✓	✓	✓	✓	Ξ	Ξ	Ξ	✓	Ξ	8 Vdc
✓	✓	✓	✓	Ξ	Ξ	Ξ	Ξ	✓	Ξ	9 Vdc
Ξ	Ξ	Ξ	Ξ	✓	✓	✓	✓	✓	Ξ	0 Vdc
Ξ	✓	✓	✓	✓	Ξ	Ξ	Ξ	Ξ	✓	-1 Vdc
✓	Ξ	✓	✓	Ξ	✓	Ξ	Ξ	Ξ	✓	-2 Vdc
Ξ	Ξ	✓	✓	✓	✓	Ξ	Ξ	Ξ	✓	-3 Vdc
✓	Ξ	Ξ	✓	Ξ	✓	✓	Ξ	Ξ	✓	-4 Vdc
Ξ	Ξ	Ξ	✓	✓	✓	✓	Ξ	Ξ	✓	-5 Vdc
✓	✓	Ξ	Ξ	✓	✓	✓	Ξ	Ξ	✓	-6 Vdc
Ξ	Ξ	✓	Ξ	✓	✓	Ξ	✓	Ξ	✓	-7 Vdc
✓	Ξ	Ξ	Ξ	Ξ	✓	✓	✓	Ξ	✓	-8 Vdc
Ξ	Ξ	Ξ	Ξ	✓	✓	✓	✓	Ξ	✓	-9 Vdc

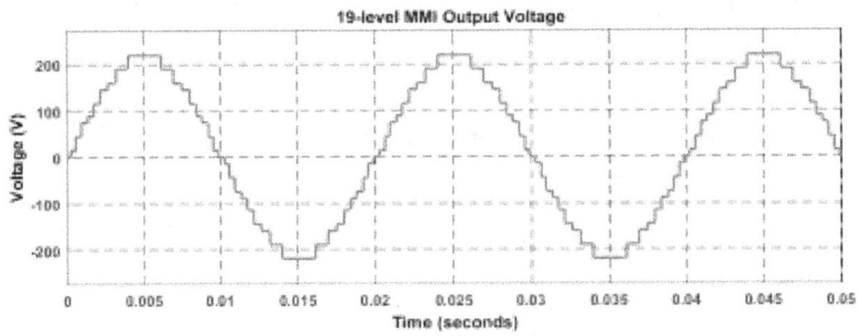

(a)

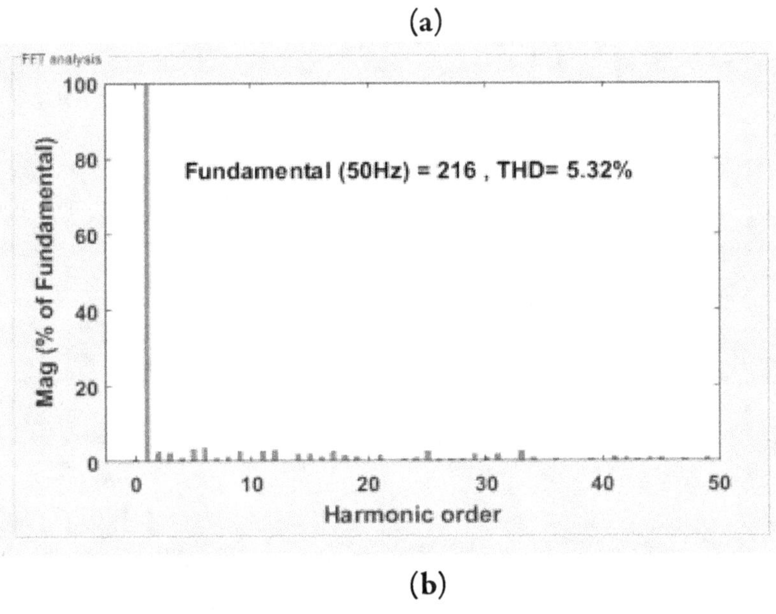

(b)

Figure 3.16 (a) 19-Level output voltage (b)FFT analysis of 19-Level Inverter

3.11.7 21-Level MMI

When the DC sources in Figure 3.2a are in the ratio of 1: 2: 3: 4 a 21-level output voltage is generated from the proposed structure. For simulation purpose the DC sources are taken as $V_{pv} = 24V$, $V_{pv2} =$

48V, $V_{pv3} = 72V$, $V_{pv4} = 96V$ (24: 48: 72: 96). Table 3.20 displays the

triggering sequence of the suggested 21-level AMMI.

Table 3.20 Triggering sequence for 21-Level AMMI

S1	S2	S3	S4	B1	B2	B3	B4	P1	P2	OUTPUT
✓	Ξ	Ξ	Ξ	Ξ	✓	✓	✓	✓	Ξ	1 Vdc
✓	✓	Ξ	Ξ	✓	Ξ	✓	✓	✓	Ξ	2 Vdc
✓	Ξ	✓	Ξ	✓	✓	Ξ	✓	✓	Ξ	3 Vdc
✓	Ξ	Ξ	✓	✓	✓	✓	Ξ	✓	Ξ	4 Vdc
✓	Ξ	Ξ	✓	Ξ	✓	✓	Ξ	✓	Ξ	5 Vdc
Ξ	✓	Ξ	✓	✓	Ξ	✓	Ξ	✓	Ξ	6 Vdc
Ξ	Ξ	✓	✓	✓	✓	Ξ	Ξ	✓	Ξ	7 Vdc
✓	Ξ	✓	✓	Ξ	✓	Ξ	Ξ	✓	Ξ	8 Vdc
Ξ	✓	✓	✓	✓	Ξ	Ξ	Ξ	✓	Ξ	9 Vdc
✓	✓	✓	✓	Ξ	Ξ	Ξ	Ξ	✓	Ξ	10 Vdc
Ξ	Ξ	Ξ	Ξ	✓	✓	✓	✓	✓	Ξ	0 Vdc
Ξ	✓	✓	✓	✓	Ξ	Ξ	Ξ	Ξ	✓	-1 Vdc

✓ Ξ ✓ ✓ Ξ	✓	Ξ Ξ	Ξ ✓	-2 Vdc
Ξ Ξ ✓ ✓ ✓	✓	Ξ Ξ	Ξ ✓	-3 Vdc
Ξ ✓ Ξ ✓ ✓	Ξ	✓ Ξ	Ξ ✓	-4 Vdc
✓ Ξ Ξ ✓ Ξ	✓	✓ Ξ	Ξ ✓	-5 Vdc
Ξ Ξ Ξ ✓ ✓	✓	✓ Ξ	Ξ ✓	-6 Vdc
Ξ Ξ ✓ Ξ ✓	✓	Ξ ✓	Ξ ✓	-7 Vdc
Ξ ✓ Ξ Ξ ✓	Ξ	✓ ✓	Ξ ✓	-8 Vdc
✓ Ξ Ξ Ξ Ξ	✓	✓ ✓	Ξ ✓	-9 Vdc
Ξ Ξ Ξ Ξ ✓	✓	✓ ✓	Ξ ✓	-10 Vdc

Figure 3.17 illustrates the simulation output of the suggested 21-level AMMI. Figure 3.17 indicates the peak output voltage and THD as

$V_{peak} = 240V$ and 5.06 %.

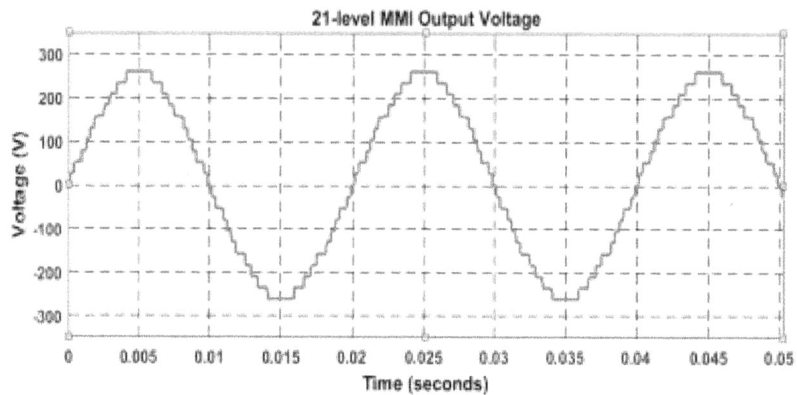

(a)

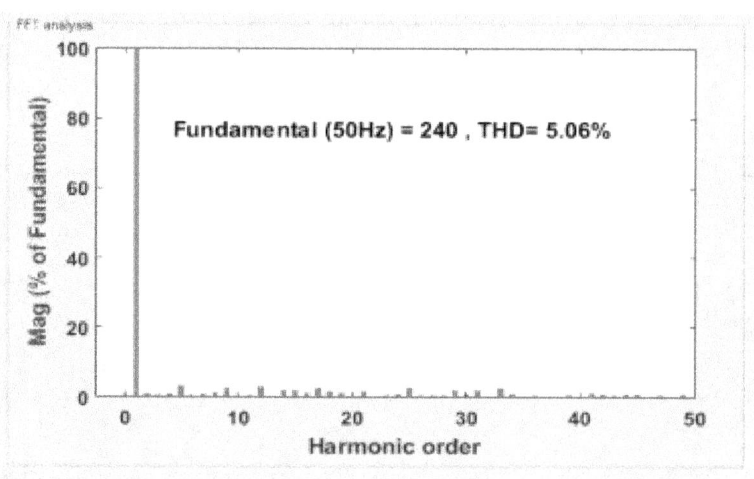

(b)

Figure 3.17 (a) 21-Level output voltage (b)FFT analysis of 21-Level Inverter

3.11.8 23-Level MMI

When the DC sources in Figure 3.2a are in the ratio of 1: 2: 3: 5 a 23-level output voltage is generated from the proposed structure. For simulation purpose the DC sources are taken as $V_{pv} = 24V$, $V_{pv2} =$

$48V$, $V_{pv3} = 72V$, $V_{pv4} = 120V$ (24: 48: 72: 120). Table 3.21 shows the

switching sequence of the proposed 23-level MMI. Figure 3.18 presents the simulation output of the proposed 23-level asymmetric multilevel inverter.

Figure 3.18 indicates the peak output voltage as $V_{peak} = 264V$. Moreover, the

THD obtained is 4.96%

Table 3.21 Triggering sequence for 23-Level AMMI

S1	S2	S3	S4	B1	B2	B3	B4	P1	P2	Output voltage
✓	Ξ	Ξ	Ξ		✓	✓	✓	✓	Ξ	1 V
Ξ	✓	Ξ	Ξ	✓	Ξ	✓	✓	✓	Ξ	2 V
Ξ	Ξ	✓	Ξ	✓	✓	X	✓	✓	Ξ	3 V
✓	Ξ	✓	Ξ	Ξ	✓	X	✓	✓	Ξ	4 V
Ξ	Ξ	Ξ	✓	✓	✓	✓	X	✓	Ξ	5 V
✓	Ξ	Ξ	✓	X	✓	✓	X	✓	Ξ	6 V
Ξ	✓	Ξ	✓	✓	Ξ	✓	X	✓	Ξ	7 V
Ξ	Ξ	✓	✓	✓	✓	X	X	✓	Ξ	8 V
✓	Ξ	✓	✓	X	✓	X	X	✓	Ξ	9 V
Ξ	✓	✓	✓	✓	X	X	X	✓	Ξ	10 V
✓	✓	✓	✓	X	X	X	X	✓	Ξ	11 V

Ξ Ξ Ξ ✓	✓	✓	✓	✓ Ξ	(
Ξ ✓ ✓ ✓ ✓	Ξ	Ξ	Ξ	X ✓	-1
✓ Ξ ✓ ✓ Ξ	✓	Ξ	Ξ	X ✓	-
Ξ Ξ ✓ ✓ ✓	✓	Ξ	Ξ	X ✓	-3
Ξ ✓ Ξ ✓ ✓	Ξ	✓	Ξ	X ✓	-4

Table 3.21 (Continued)

S1 S2 S3 S4	B1	B2	B3	B4	P1 P2	Output voltage
✓ Ξ Ξ ✓ Ξ	✓	✓	Ξ		X ✓	-5 Vd
Ξ Ξ Ξ ✓ ✓	✓	✓	Ξ		X ✓	-6 Vd
✓ Ξ ✓ Ξ X	✓	Ξ	✓		X ✓	-7 Vd
X Ξ ✓ Ξ ✓	✓	Ξ	✓		X ✓	-8 Vd
Ξ ✓ X Ξ ✓	X	Ξ	✓		X ✓	-9Vd
✓ X Ξ Ξ ✓	Ξ	✓	✓		X ✓	-10V
Ξ Ξ Ξ Ξ ✓	✓	✓	✓		X ✓	-11V

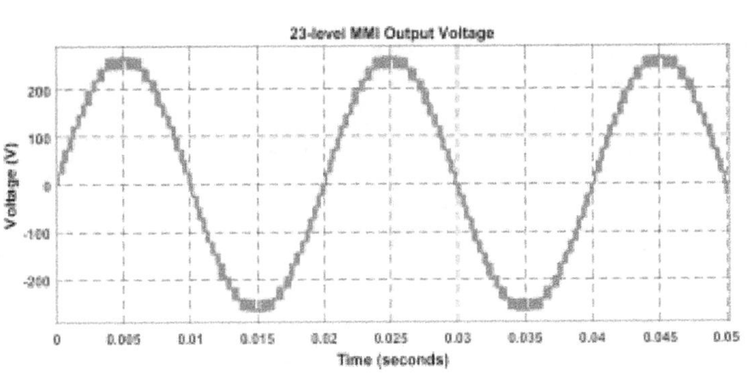

(a)

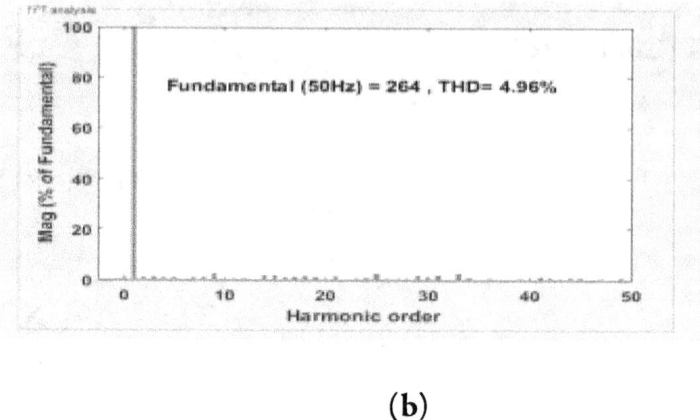

(b)

Figure 3.18 (a) 23-Level output voltage (b)FFT analysis of 23-Level Inverter

3.11.9 25-Level MMI

When the DC sources in Figure 3.2a are in the ratio of 1: 2: 3: 6 a 25-level output voltage is generated from the proposed structure. For simulation purpose the DC sources are taken as $V_{pv} = 24V$, $V_{pv2} =$

48V, $V_{pv3} = 72V$, $V_{pv4} = 144V$ (24: 48: 72: 144) . Table 3.22 shows the

switching sequence of the proposed 25-level MMI. Figure 3.19 presents the simulation output of the proposed 25-level asymmetric multilevel inverter.

Figure 3.19 clearly indicates the peak output voltage as $V_{peak} =$

288V.Moreover, the THD obtained is 4.51%

Table 3.22 Triggering sequence for 25-Level AMMI

S1	S2	S3	S4	B1	B2	B3	B4	P1	P2	Output voltage
✓	Ξ	Ξ	Ξ	Ξ	✓	✓	✓	✓	Ξ	1 Vdc
Ξ	✓	Ξ	Ξ	✓	Ξ	✓	✓	✓	Ξ	2 Vdc
Ξ	Ξ	✓	Ξ	✓	✓	X	✓	✓	Ξ	3 Vdc
✓	Ξ	✓	Ξ	Ξ	✓	X	✓	✓	Ξ	4 Vdc
Ξ	✓	✓	Ξ	✓	X	X	✓	✓	Ξ	5 Vdc
Ξ	Ξ	Ξ	✓	✓	✓	✓	X	✓	Ξ	6 Vdc
✓	Ξ	Ξ	✓	X	✓	✓	X	✓	Ξ	7 Vdc
Ξ	✓	Ξ	✓	✓	Ξ	✓	X	✓	Ξ	8 Vdc
Ξ	Ξ	✓	✓	✓	✓	X	X	✓	Ξ	9 Vdc
✓	Ξ	✓	✓	X	✓	X	X	✓	Ξ	10 Vdc
Ξ	✓	✓	✓	✓	X	X	X	✓	Ξ	11 Vdc
✓	✓	✓	✓	X	X	X	X	✓	Ξ	12 Vdc

Table 3.22 (Continued)

S1	S2	S3	S4	B1	B2	B3	B4	P1	P2	Output voltage
Ξ	Ξ	Ξ	Ξ	✓	✓	✓	✓	✓	Ξ	0 Vdc
Ξ	✓	✓	✓	✓	Ξ	Ξ	Ξ	X	✓	-1 Vdc
✓	Ξ	✓	✓	Ξ	✓	Ξ	Ξ	X	✓	-2Vdc
Ξ	Ξ	✓	✓	✓	✓	Ξ	Ξ	X	✓	-3 Vdc
Ξ	✓	Ξ	✓	✓	Ξ	✓	Ξ	X	✓	-4 Vdc
✓	Ξ	Ξ	✓	Ξ	✓	✓	Ξ	X	✓	-5 Vdc
Ξ	Ξ	Ξ	✓	✓	✓	✓	Ξ	X	✓	-6 Vdc
Ξ	✓	✓	Ξ	✓	Ξ	Ξ	✓	X	✓	-7 Vdc
✓	Ξ	✓	Ξ	Ξ	✓	Ξ	✓	X	✓	-8 Vdc
Ξ	Ξ	✓	Ξ	✓	✓	Ξ	✓	X	✓	-9Vdc
Ξ	✓	Ξ	Ξ	✓	Ξ	✓	✓	X	✓	-10Vdc
✓	Ξ	Ξ	Ξ	Ξ	✓	✓	✓	X	✓	-11Vdc
Ξ	Ξ	Ξ	Ξ	✓	✓	✓	✓	X	✓	-12Vdc

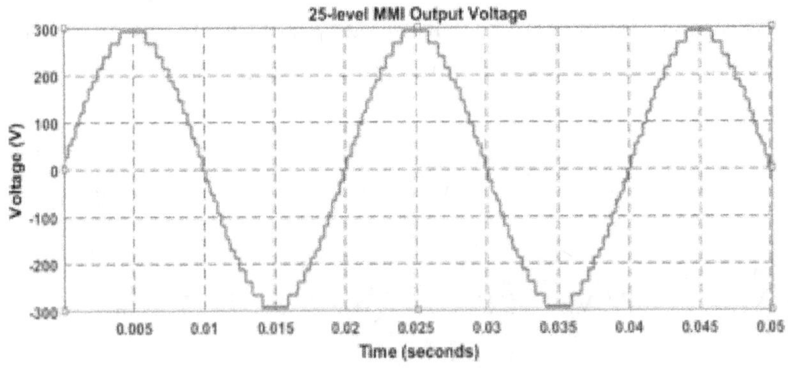

(a)

Figure 3.19 (Continued)

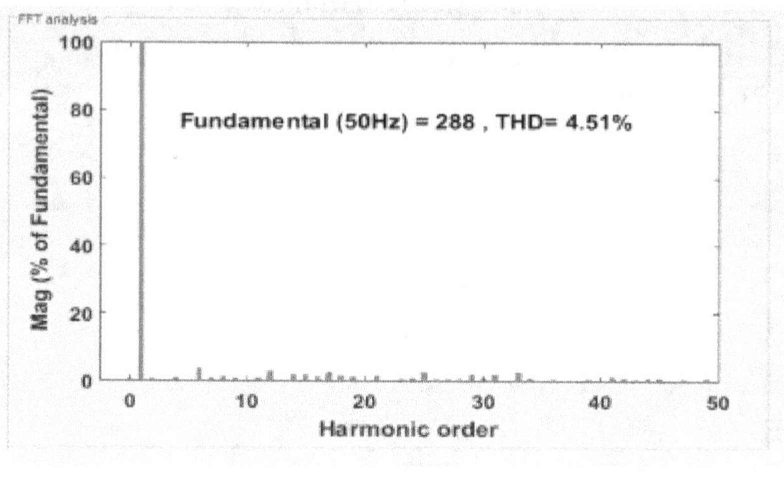

(b)

Figure 3.19 (a) 25-Level output voltage (b)FFT analysis of 25-Level Inverter

3.11.10 27-Level MMI

When the PV inputs for the circuit shown in Figure 3.2a are in the ratio of 1: 2: 3: 7 a 27-level output is produced from the suggested structure.

For experimental and simulation purpose the DC sources are taken as $V_{pv} =$

$24V, V_{pv2} = 48V, V_{pv3} = 72V, V_{pv4} = 168V$ (24: 48: 72: 168).

Table 3.5 shows the triggering sequence of the suggested 27-level AMMI. Figure 3.20 illustrates the output of the suggested 27-level AMMI.

Figure 3.20 clearly indicates the peak output voltage as $V_{peak} =$

312V.Moreover, the THD obtained is 3.62 % which is very less and satisfies the IEEE 519 standards.

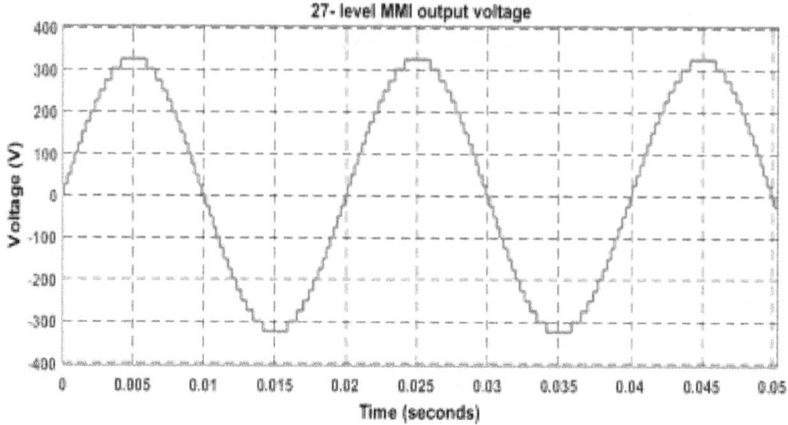

(a)

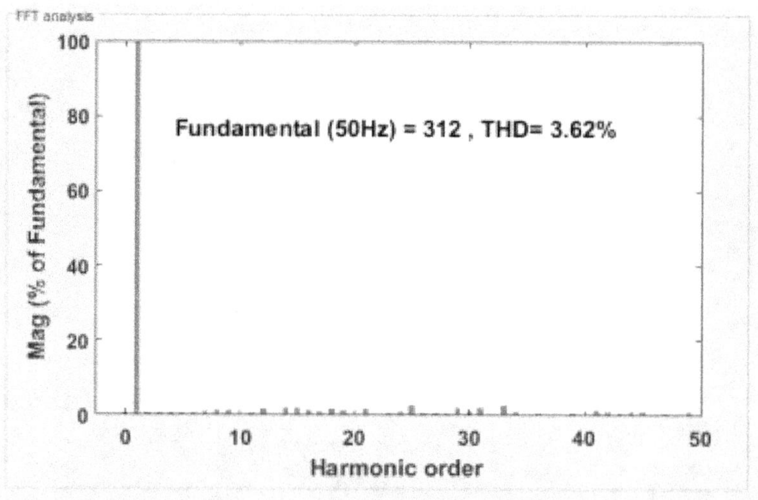

(b)

Figure 3.20 (a) 27-Level output voltage (b)FFT analysis of 27-Level Inverter

3.11.11 29-Level MMI

When the inputs of Figure 3.2a are in the ratio of 1: 2: 4: 7 a 29- level output is produced from the proposed structure. For simulation purpose the DC sources are taken as $V_{pv} = 24V$, $V_{pv2} = 48V$, $V_{pv3} = 96V$, $V_{pv4} =$

168V (24: 48: 96: 168).Table 3.23 shows the triggering sequence of the

suggested 29-level AMMI.

Table 3.23 Triggering Sequence for 29-Level AMMI

S1	S2	S3	S4	B1	B2	B3	B4	P1	P2	Output voltage
✓	Ξ	Ξ	Ξ	Ξ	✓	✓	✓	✓	Ξ	1 Vdc
Ξ	✓	Ξ	Ξ	✓	Ξ	✓	✓	✓	Ξ	2 Vdc
✓	✓	Ξ	Ξ	Ξ	Ξ	✓	✓	✓	Ξ	3 Vdc
Ξ	Ξ	✓	Ξ	✓	✓	X	✓	✓	Ξ	4 Vdc
✓	Ξ	✓	Ξ	Ξ	✓	X	✓	✓	Ξ	5 Vdc
Ξ	✓	✓	Ξ	✓	X	X	✓	✓	Ξ	6 Vdc
Ξ	Ξ	Ξ	✓	✓	✓	✓	X	✓	Ξ	7 Vdc
✓	Ξ	Ξ	✓	X	✓	✓	X	✓	Ξ	8 Vdc
Ξ	✓	Ξ	✓	✓	Ξ	✓	X	✓	Ξ	9 Vdc
✓	✓	Ξ	✓	Ξ	Ξ	✓	Ξ	✓	Ξ	10 Vdc
Ξ	Ξ	✓	✓	✓	✓	X	X	✓	Ξ	11 Vdc

✓	Ξ	✓	✓	X	✓	X	X	✓	Ξ

12 Vdc

Ξ	✓	✓	✓	✓	X	X	X	✓	Ξ

13 Vdc

✓	✓	✓	✓	X	X	X	X	✓	Ξ

14 Vdc

Ξ	Ξ	Ξ	Ξ	✓	✓	✓	✓	✓	Ξ

0 Vdc

Ξ	✓	✓	✓	✓	Ξ	Ξ	X	X	✓

-1 Vdc

Table 3.23 (Continued)

S1	S2	S3	S4	B1	B2	B3	B4	P1	P2	Output voltage
✓	Ξ	✓	✓	Ξ	✓	Ξ	Ξ	X	✓	-2Vdc
Ξ	Ξ	✓	✓	✓	✓	Ξ	Ξ	X	✓	-3 Vdc
✓	✓	Ξ	✓	Ξ	Ξ	✓	Ξ	X	✓	-4 Vdc
Ξ	✓	Ξ	✓	✓	Ξ	✓	Ξ	X	✓	-5Vdc
✓	Ξ	Ξ	✓	Ξ	✓	✓	Ξ	X	✓	-6 Vdc
Ξ	Ξ	Ξ	✓	✓	✓	✓	Ξ	X	✓	-7Vdc
Ξ	✓	✓	Ξ	✓	Ξ	Ξ	✓	X	✓	-8 Vdc
✓	Ξ	✓	Ξ	Ξ	✓	Ξ	✓	X	✓	-9 Vdc
Ξ	Ξ	✓	Ξ	✓	✓	Ξ	✓	X	✓	-10Vdc
✓	✓	Ξ	Ξ	Ξ	Ξ	✓	✓	X	✓	-11Vdc
Ξ	✓	Ξ	Ξ	✓	Ξ	✓	✓	X	✓	-12Vdc
✓	Ξ	Ξ	Ξ	Ξ	✓	✓	✓	Ξ	✓	-13Vdc
Ξ	Ξ	Ξ	Ξ	✓	✓	✓	✓	X	✓	-14Vdc

Figure 3.21 illustrates the output of the suggested 29 level AMMI. Figure 3.21 clearly indicates the peak output voltage as $V_{peak} =$ 336V.Moreover, the THD obtained is 2.12 % and satisfies the IEEE 519

standards.

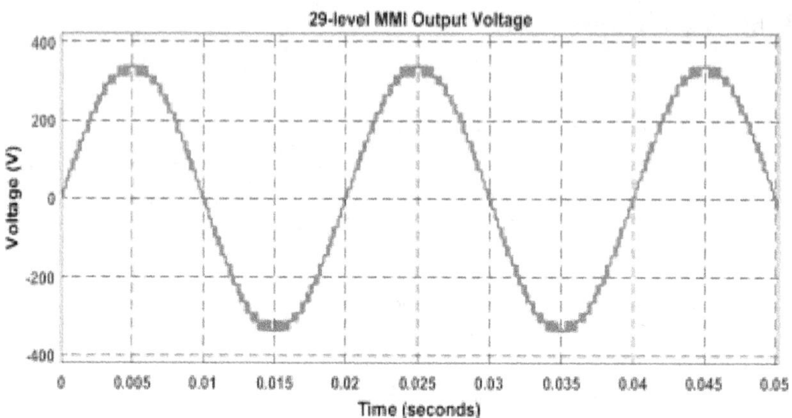

(a)

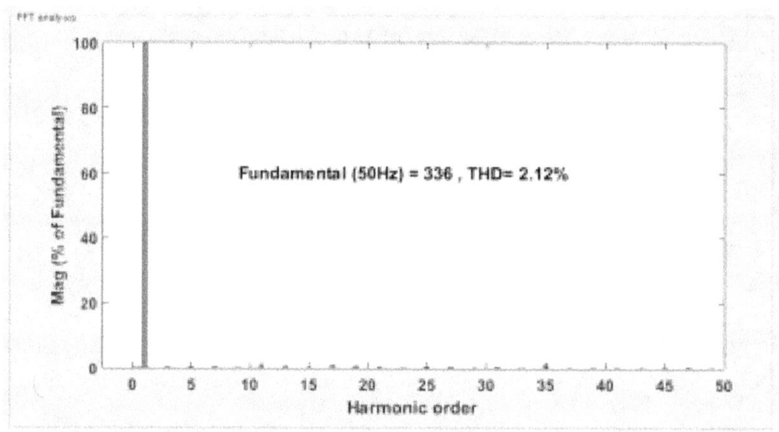

(b)

Figure 3.21 (a) 29-Level output voltage (b)FFT analysis of 29-Level Inverter

3.11.12 31-Level MMI

When the inputs of Figure 3.1 are in 1: 2: 4: 8 ratio a 31-level output is produced from the proposed structure. For experimental and simulation purpose the DC sources are taken as $V_{pv} = 24V$, $V_{pv2} =$

48V, V_{pv3} = 96V, V_{pv} = 192V (24: 48: 96: 192). Table 3.9 shows the

triggering sequence of the suggested 31-level AMMI. Figure 3.22 illustrates the output of the suggested 31-level AMMI. Figure 3.22 clearly indicates the

peak output voltage as V_{peak} = 360V.Moreover, the THD obtained is 1.76%
and satisfies the IEEE 519 standards.

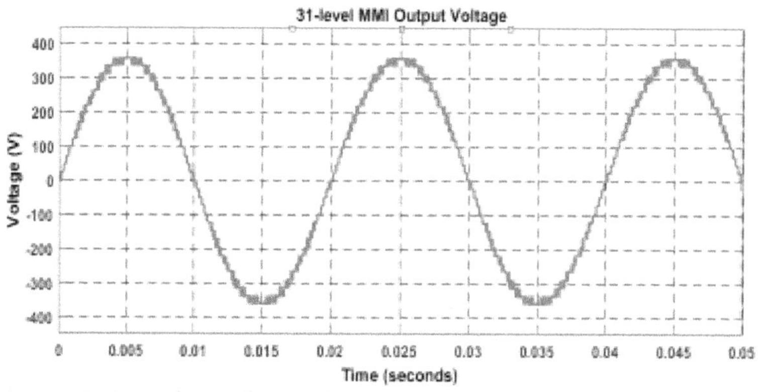

(a)

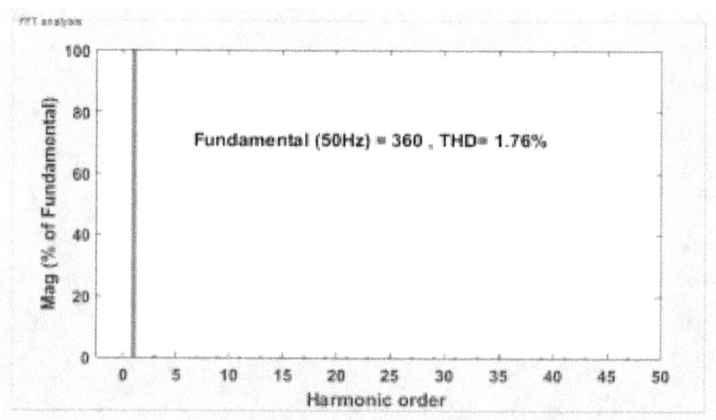

(b)

Figure 3.22 (a) 31-Level output voltage (b)FFT analysis of 31-Level Inverter

EXPERIMENTAL RESULTS

The functional block diagram representation of the MMI is presented in Figure 3.23. To produce the gate pulses, the advanced digital controller i.e. field-programmable gate array (FPGA) is used which computes faster when compared to the microcontroller-based pulse generators. The authority of proposed symmetric topology is implementing a laboratory-based model, and the experimental test has been conducted. For the prototype, the

input source value is 50V, this voltage is obtained by connecting four 12V, 7A

PV panel in series and the remaining 2V is boosted by the boost converter. The load value is chosen as R=100Ω and L=10mH respectively.

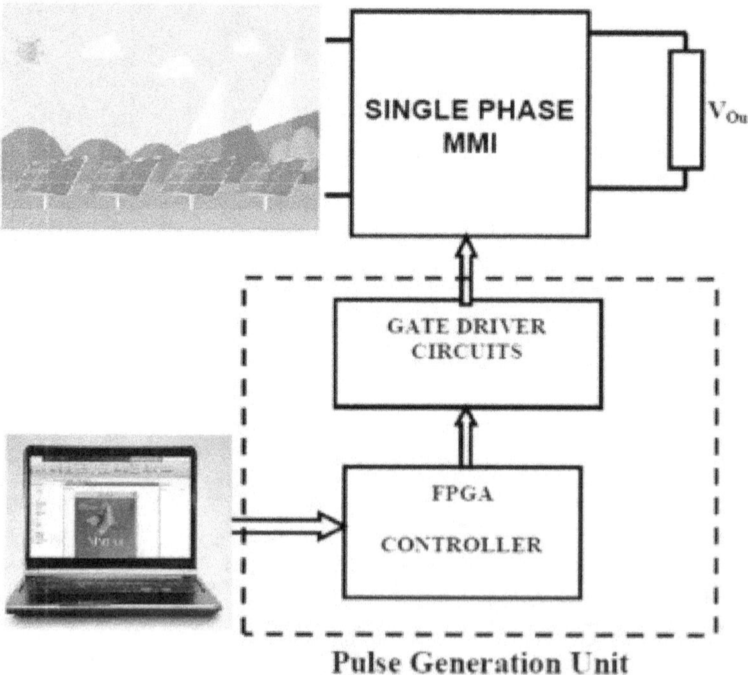

Figure 3.23 Block Diagram Representation for Hardware Implementation

9-level Symmetric Modular Multilevel Inverter

In this part of the research, a model of FPGA based 9-level SMMI has been made-up to improve the power quality. Figure 3.24 shows the 1kWp solar PV plant. Figure 3.25 displays the experimental setup of the suggested MMI.

Figure 3.24 1kWp Solar PV Plant

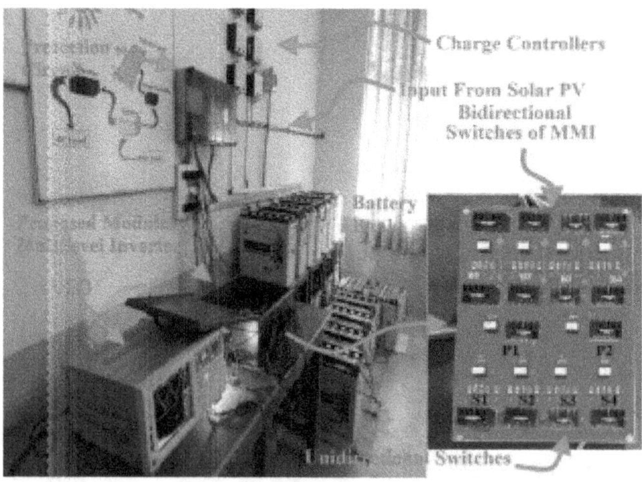

Figure 3.25 Prototype of the Proposed MMI

The suggested MMI includes four DC inputs, FGA25N120 IGBT switching elements, TLP250 driver circuit, FPGA controller, RL load of

100 fl & 10 mH and connecting probes. The switching frequency is 1kHz.

Substantial development in power quality is succeeded by lessening the harmonic content in the output using a selective harmonic elimination technique. A 9-level SMMI is fabricated to reduce the harmonics of the output voltage.

The selective harmonic elimination method is implemented in the FPGA controller and produces the appropriate switching pulses to the inverter through gate driver circuits. For the 9-level SMMI four nonlinear equations

are solved to produce four triggering angles (θ_1, θ_2, θ_3, θ_4). These triggering

angles are saved as lookup table in the memory of FPGA controller. In Matlab the switching angles are calculated and stored in the lookup table for different modulation index. Using the hardware description language (HDL) coder in Matlab the switching angles generation code which is in .m format is converted to .vhd code. This conversion is necessary since the FPGA kit works with .vhd code. Therefore, the converted .vhd code is run in the controller.

Thus, the FPGA controller generates the pulses as the output. These pulses are used to trigger the switches of the SMMI. By triggering the switches a 9-level output voltage is obtained in which the 3_{rd}, 5_{th} and 7_{th} order harmonics are minimised. Figure 3.26 displays the 9-level output voltage and its corresponding output current waveforms. The measured maximum output voltage at load ($V_{Out(rms)}$) is 141.42V with a 9-level stepped waveform, and

the corresponding output current value is 4.94A. To authorize the suggested

SMMI structure, the output harmonic spectrum is also revealed in Figure 3.26.

Tek Trig'd M Pos: 40.20ms
PEAK VOLTAGE:140V
VOLTAGE:70V/DIV
CURRENT: 10 A/DIV
Time (Seconds)

(a)

Tek Trig'd M Pos: 36.00ms
CH1
V RMS 141.1V THD-F 0.14%
THD-R 0.14%
Harmonic Fnd
Freq 50.1Hz %Fund 100.0%
hRMS 141.1V Φ 0.00°
Fnd 3 5 7 9 11 13 15 17 19 21 23 25
M 10.0ms

Tek Trig'd M Pos: 10.10ms
CH3
I RMS 4.94A THD-F 7.22%
THD-R 7.22%
Harmonic Fnd
Freq 50.0Hz %Fund 100.0%
hRMS 4.94A Φ 0.00°
Fnd 3 5 7 9 11 13 15 17 19 21 23 25
M 10.0ms

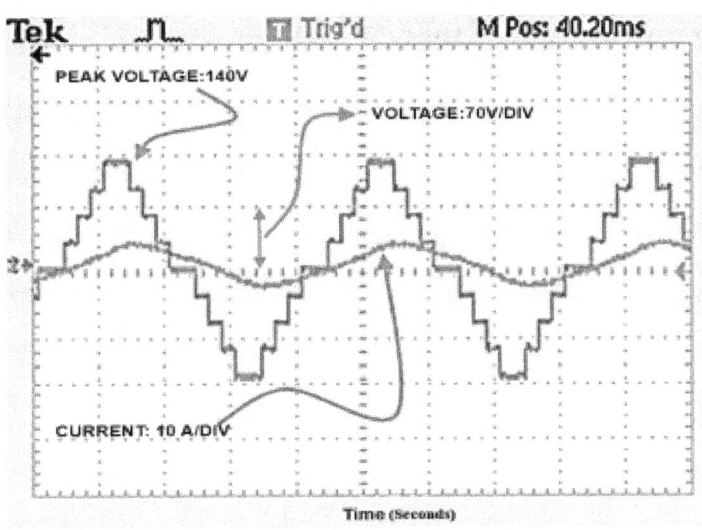

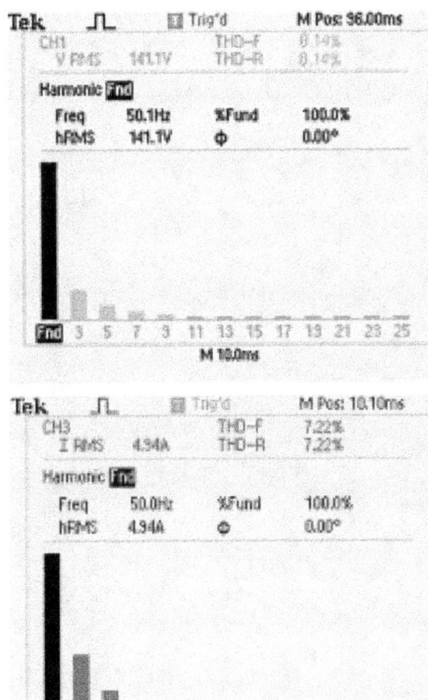

(b) (c)

Figure 3.26 Output Voltage and Current Waveform of 9- Level Inverter with Voltage and Current THDs

In the experimental configuration power quality analyzer is used to calculate the Total Harmonic Distortion (THD). The measurement from the power quality analyzer reveals that the voltage and current waveform THDs are 8.14% and 7.22% respectively. This is similar to both the Simulink model's simulation results.

27-level Asymmetric Modular Multilevel Inverter

For experimental purpose, the DC sources are taken as $V_{DC} = 24V$, $V_{DC2} = 48V$, $V_{DC3} = 72V$, $V_{DC4} = 168V$ (24: 48: 72: 168). Figure 3.27

displays the appropriate stepped output voltage, output current and THDs. The measured peak inverter voltage at load ($V_{Out(rms)}$) is 220.61V with 27-level

stepped waveform. The measurement from power quality analyzer reveals that the voltage and current waveform THDs are 6.17% and 5.13% respectively

31-level Asymmetric Modular Multilevel Inverter

For experimental and simulation purpose the DC sources are taken as $V_{DC} = 24V$, $V_{DC2} = 48V$, $V_{DC3} = 96V$, $V_{DC4} = 192V$ (24: 48: 96: 192).

Figure 3.28 displays the experimental output of the suggested 31 AMMI with its respective THDs. The measured maximum output voltage at load

($V_{Out(rms)}$) is 254.55V with a 31-level stepped waveform. The measurement from the power quality analyzer reveals that the THDs of voltage and current are 3.17% and 2.33% respectively.

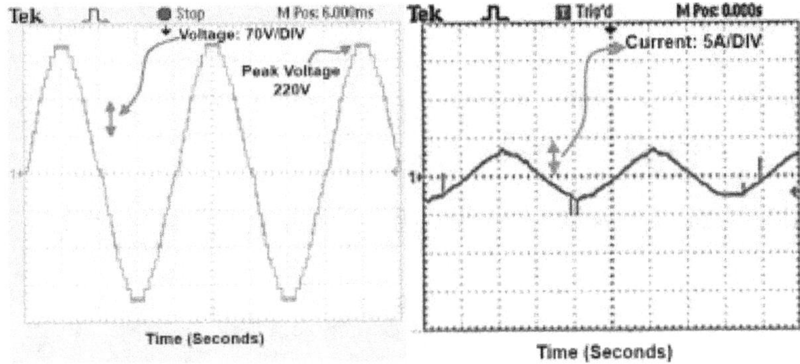

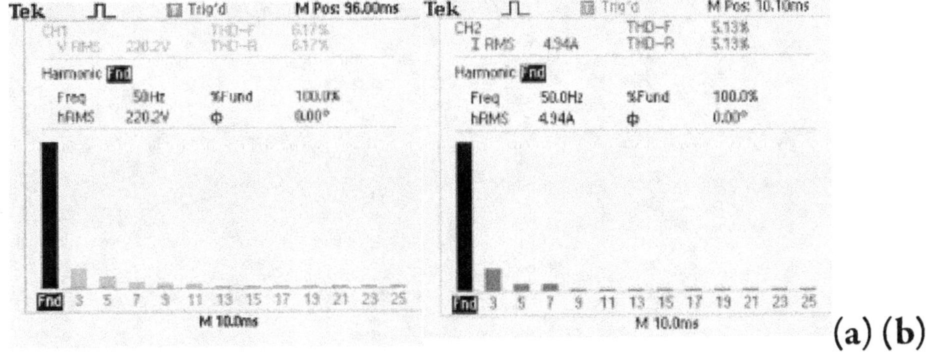

Tek
Stop
M Pos: 6.000ms
Voltage: 70V/DIV
Peak Voltage
220V
Time (Seconds)
Tek
Trig'd
M Pos: 0.000s
Current: 5A/DIV
Time (Seconds)

Tek
Trig'd
M Pos: 96.00ms
CH1
THD-F
6.17%
V RMS
220.2V
THD-R
6.17%
Harmonic Fnd
Freq
50Hz
%Fund
100.0%
hRMS
220.2V
Φ
0.00°
Fnd 3 5 7 9 11 13 15 17 19 21 23 25
M 10.0ms
Tek
Trig'd
M Pos: 10.10ms
CH2
THD-F
5.13%
I RMS
4.94A
THD-R
5.13%
Harmonic Fnd
Freq
50.0Hz
%Fund
100.0%
hRMS
4.94A
Φ
0.00°
Fnd 3 5 7 9 11 13 15 17 19 21 23 25
M 10.0ms

(a) (b)

a. (d)

Figure 3.27 Output Voltage and Current Waveform of 27-Level Inverter with Voltage and Current THDs

Table 3.24 Comparison of Results for the Proposed MMI

S. No.	Level	THD (%)	
		Simulation	Experimental
1	9	7.54	8.14
2	27	3.62	6.17
3	31	1.76	3.17

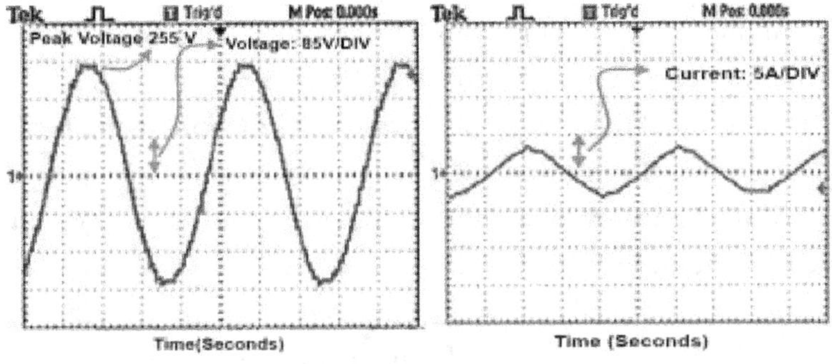

a.

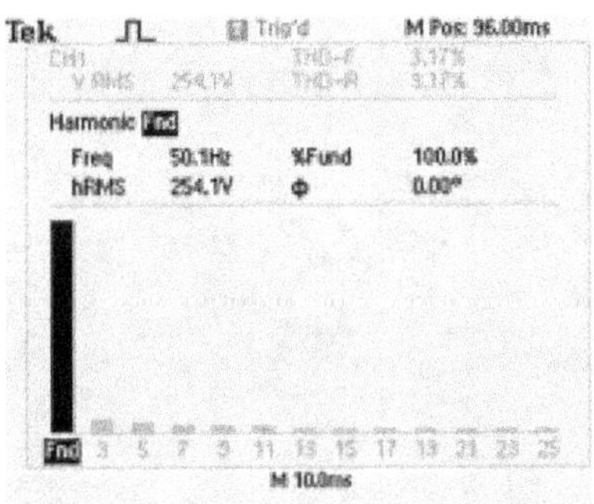

(b)

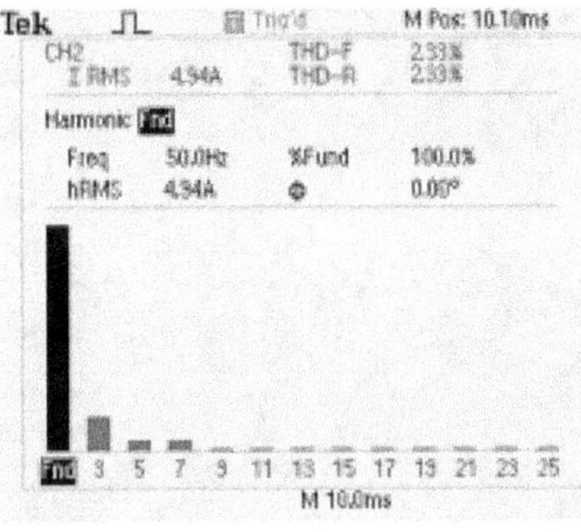

(c) (d)

Figure 3.28 Output Voltage and Current Waveform of 31-Level Inverter with Current and Voltage THDs

SUMMARY

The proposed modular multilevel inverters for both symmetric and asymmetric topologies are discussed. The generalized mathematical expression for the number of levels, number of switches, and blocking voltage are presented for both symmetric 9-level inverter and asymmetric 27,31-level inverter. The proposed SHE PWM is employed in both simulation and experimental tests. Experimental setup designed for 9-level,27-level, and 31-level structures are analysed and compared with the simulation outputs with respect to setup and other existing topologies.

CHAPTER 4

HARMONIC ELIMINATION IN MODULAR MULTILEVEL INVERTER USING OPTIMIZATION TECHNIQUES

INTRODUCTION

Multilevel Inverters (MLI) have several benefits and different configurations. Unfortunately, owing to their poor THD, most of the present inverters cannot achieve good output waveforms. This is because the switching angles are not organized correctly for each level. To achieve good power quality, the switching angles must be properly arranged to obtain the lowest THD. Harmonics are characterized as the integral multiples of the fundamental frequency and occur as the voltage or current distortion of the waveform. The existence of harmonics in the inverter output can contribute to overheating in neutral conductors of machines, erratic operation of circuit breakers, malfunctioning of microprocessor-based equipment, degradation of power factor correction capacitors, and pronounced magnetic fields near switchgears and transformers. In order to remove unnecessary harmonics in MLI with equivalent DC sources, Kumar *et al.* (2019) proposed various modulation techniques.

Many of the PWM methods typically use a carrier signal to compare with the reference signal and the obtained square pulses are used to trigger the switches of MLI. The triggering pulses are created from the modulating signal and the reference signal. The switching frequency is

equivalent to the carrier frequency and is designated as f_c. The carrier frequency should be a lower value to produce the output voltage with minimum switching losses which are caused by the semiconductor switches. Moreover it is not advisable to use modulation techniques, like carrier-based PWM or Space- Vector Modulation (SVM), below 1 kHz level, because the side bands across the carrier frequency act as lower order harmonics, resulting in a greater distortion of the load voltage and current.

To reduce the switching losses a variety of pre-programmed low-frequency PWM techniques are implemented. These methods provide offline solutions which ensures the required fundamental voltage while removing the lower order harmonics. SHE method eliminates different order harmonics in the inverter voltage by choosing appropriate switching angles. These switching angles are obtained by solving a set of complex transcendental equations. The mathematical solution to nonlinear transcendental equations, including trigonometric terms, are very difficult to obtain. This is a fundamental problem in this method. Mahato *et al.* (2019) have put forth a generalized approach for SHE in MLI.

Various iterative approaches are implemented to compute the nonlinear equations. Rai & Chakravorty (2019) have taken up the drawbacks of iterative approaches such as Newton Raphson because they rely on initial assumptions and divergence problem for several inverter stages. The resulting theory transforms transcendental equations to an analogous series of polynomial equations and then uses the statistical principle of the resultant to consider all the solutions available for this comparable problem. The resultant theory is restricted to 6 triggering angles for equivalent DC inputs and 3 triggering angles for dissimilar DC inputs. Furthermore, when level or DC voltage varies, a new expression is evolved which becomes complicated to solve and is time consuming. The step modulation proposed by

Liu *et al.* (2009) computes the triggering angles for SMLI and it does not extend this method for ASMLI.

PROBLEM STATEMENT

Applying dissimilar DC sources contributes to a variety of high-degree equations, that are difficult to be solved. This restriction is remedied by using different methods of optimization techniques such as BEE Colony Optimization (BCO),Genetic Algorithm (GA), and Particle Swarm Optimization (PSO). For the suggested technique, the objective function is developed and corresponding operations with regard to GA, PSO, and BA are conducted, to achieve the optimal triggering angles required for the inverters. An experimental setup for the method which provides the least THD is implemented for a 1kWp solar PV plant.

4.2.1 Representation of 9-level Output Staircase Waveform in Fourier Series

The selective harmonic elimination method is very appropriate for MMIs. By implementing this method, the selected harmonics ($3_{rd}, 5_{th}$ and 7_{th}) in the output of the 9-level MMI is drastically reduced. Therefore, the use of filters at the end of the MMI is also reduced. The selected harmonics can be removed from the staircase waveform by implementing additional switching within the staircase waveform. During every half cycle of the 9-level output, the semiconductor switches are switched on and off continuously and these switching's are distributed equally over each cycle with quarter wave symmetry. In a staircase waveform if the waveform repeats the same pattern for every quarter cycle it is called quarter wave symmetry. Figure 4.1 illustrates the output staircase waveform of the 9-level SMMI. The PV inputs

V_{PV1} to V_{PV4} presented in Figure 4.1 are obtained from PV sources. The output of MMI is given in Equation (4.1).

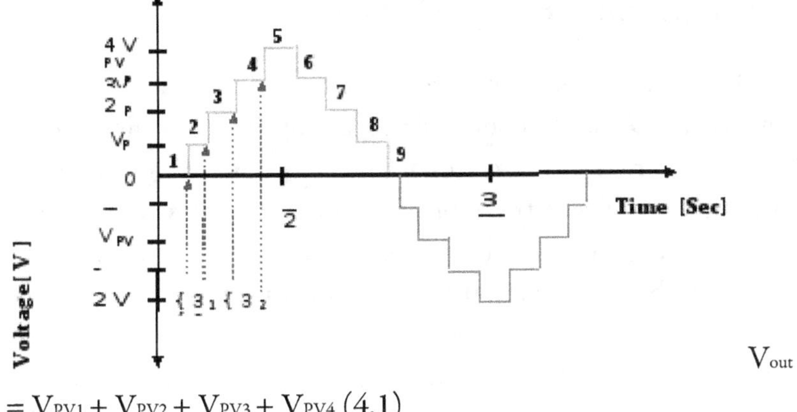

$$V_{out} = V_{PV1} + V_{PV2} + V_{PV3} + V_{PV4} \quad (4.1)$$

Figure 4.1 Output of 9-level MMI

The harmonic components from the 9-level staircase waveform can be extracted using Fourier series. In Fourier series any periodic function can be represented by an infinite addition of harmonically related trigonometric functions. By executing the Fourier series, the 9-level waveform which is non-sinusoidal in nature can be represented in equations with sine or cosine terms. Thus, by simplifying the Fourier series the fundamental and harmonic equations are derived. Therefore Equation 4.2 represents the Fourier series of a periodic function in the mathematical form.

CX) CX)

$$F(t) = A_O + L\ A_n \cos(nwt) + L\ B_n \sin(nwt)$$

(4.2)

n=1 n=1

here

2TI

4

$$_A_O = TI\ J\ F(t)\ dwt \quad (4.3)$$

O

$$A_n = \frac{4}{\pi}\int_0^{\frac{\pi}{2}} F(t)\cos nwt \, dwt \quad (4.4)$$

$$B_n = \frac{4}{\pi}\int_0^{\frac{\pi}{2}} F(t)\sin nwt \, dwt \quad (4.5)$$

A0, An, and Bn are termed as the Fourier elements, the angular frequency is w ,(w = 2πf)

The half wave symmetry and the quarter wave symmetry are used to simplify the Fourier series. In a staircase waveform if each half cycle is the mirror image of the next half cycle then it is called half wave symmetry. The halfwave symmetry eliminates the even harmonics from the output waveform. The presence of even order harmonics leads to resonance in the network. In Equation (4.2), due to the half wave symmetry and the quarter-wave

symmetry the Fourier elements, A0 , An and the integral multiples of

fundamental (even harmonics) are equated to zero. Moreover, to establish the amplitude of the Fourier elements (B_n) of odd harmonics (1/4)th of a fundamental period is integrated. From Equation (4.5),

$$B_n = \frac{4}{\pi}\int_0^{\frac{\pi}{2}} F(t)\sin nwt \, dwt$$

Where $F(t) = V_{out}$

2TI

4

$$\_B_n = \mathrm{TI} \int_o V_{out} \sin nwt \, dwt \quad (4.6)$$

For all n, the Fourier series is given as

$$F(t)$$

$$^{\infty}$$

$$= L\ B_n \sin (nwt)$$

$$n=1$$

$$(4.7)$$

Substituting θ = wt in Equation (4.5),

$$\frac{4}{2TI}$$

$$\_B_n = TI \int F(t) \sin n\theta\ d\theta \quad (4.8)$$

$$O$$

4V r

$$TI\ 2$$

$$TI\ 2$$

$$TI\ 2$$

$$TI$$

$$2\ 1$$

$$B = {}^{PV} I \int \sin n\theta\ d\theta + \int \sin n\theta\ d\theta + \int \sin n\theta\ d\theta + \int \sin n\theta\ d\theta\ I$$

$$(4.9)$$

$$n\ TI$$

I

I

$2\ 3\ J$

B_n

T

$\underline{4V_{PV}}$

$= L \cos(n\blacklozenge)\ (4.10)$

$n\overline{T}I$

k

k=1

$$B = \underline{4V_{PV}}[\cos(n\blacklozenge) + \cos(n\blacklozenge) + \cos(n\blacklozenge) + \cos(n\blacklozenge)]\ (4.11)$$
$n\ n\overline{T}I\ 1\ 2\ 3\ 4$

Where n is the harmonic order, T is the number of switching angles each quarter and is also equal to number of PV sources, $\blacklozenge_k$ is the k_{th} switching angle and V_{PV} indicates the DC voltages.

From Equation (4.8) to (4.11) the harmonic elements can be summarized as follows:

- The peak value of the DC component is equated to zero

- The magnitude of all even harmonics is equated to zero

• The peak value of the fundamental component and the odd harmonic component is given by,

$$V = \frac{4V_{PV}}{\pi} \sum_{k=1}^{T} \cos(\theta_k) \quad (4.12)$$

$$V = \frac{4V_{PV}}{n\pi} \sum_{k=1}^{T} \cos(n\theta_k) \quad (4.13)$$

The SHE method generates the triggering angles θ_1, θ_2, θ_3, and θ_4

for the waveform displayed in Figure 4.1. To determine triggering angles θ_1,

θ_2, θ_3, and θ_4, Equations (4.14) to (4.17) derived from Equation (4.13) ought

to be evaluated. Since these equations are indefinite and constitute trigonometric parameters, they are not possible to be solved with algebraic techniques. This contributes to the necessity of optimization methods to render the triggering angles (θ_1, θ_2, θ_3, θ_4) for producing prefered output with

reduced harmonic distortions.

The nonlinear equation of the fundamental component is written as:

$$V_1 = \frac{4V_{PV}}{\pi T}[\cos(\theta_1) + \cos(\theta_2) + \cos(\theta_3) + \cos(\theta_4)] \quad (4.14)$$

The nonlinear equations for the odd harmonic component are written as:

$$V_3 = \frac{4V_{PV}}{3\pi T}[\cos(3\theta_1) + \cos(3\theta_2) + \cos(3\theta_3) + \cos(3\theta_4)] \quad (4.15)$$

$$V_5 = \frac{4V_{PV}}{5\pi T}[\cos(5\theta_1) + \cos(5\theta_2) + \cos(5\theta_3) + \cos(5\theta_4)] \quad (4.16)$$

$$V_7 = \frac{4V_{PV}}{7\pi T}[\cos(7\theta_1) + \cos(7\theta_2) + \cos(7\theta_3) + \cos(7\theta_4)] \quad (4.17)$$

234

Four triggering angles, specifically θ_1, θ_2, θ_3, and θ_4, must be

known to obtain the desired output from four input PV sources. By solving four nonlinear equations the four triggering angles are obtained. These obtained triggering angles are stored in the lookup table and are given as pulses to trigger the switches of SMMI. By triggering the switches a 9-level output is generated in which the 3rd, 5th and 7th order harmonics are minimised.

Most notably, the solutions obtained ought to be less than $\frac{\pi}{2}$. The peak values

of the components present on the right-hand side of the Equations (4.14) to (4.17) can be regulated. The basic voltage equation presented in

Equation (4.14) can be expressed in terms of M_i as follows,

$$\cos(\theta_1) + \cos(\theta_2) + \cos(\theta_3) + \cos(\theta_4) = H * M_i \quad (4.18)$$

To remove the 3rd, 5th, and 7th order harmonics from the output Equations (4.19) to (4.22) should be solved.

$$\cos(\theta_1) + \cos(\theta_2) + \cos(\theta_3) + \cos(\theta_4) - (4 * M_i) = 0 \quad (4.19)$$
$$\cos(3\theta_1) + \cos(3\theta_2) + \cos(3\theta_3) + \cos(3\theta_4) = 0 \quad (4.20)$$
$$\cos(5\theta_1) + \cos(5\theta_2) + \cos(5\theta_3) + \cos(5) = 0 \quad (4.21)$$
$$\cos(7\theta_1) + \cos(7\theta_2) + \cos(7\theta_3) + \cos(7\theta_4) = 0 \quad (4.22)$$

Where, $M = \frac{TIV}{} = \frac{TIV}{}$

i Vmax 4* *V

H denotes the number of PV sources, V_{max} is the peak output voltage of MMI, V_{PV} is the input PV source of MMI

In the suggested MMI based PV system the PV-cell voltages vary from 36V to 48V, and from the battery-driven energy storage network centered on MMI, the battery voltages often shift according to their state of charge (SoC). The various voltages that enter the inverter stages would, therefore, trigger fundamental non-regulated harmonics of higher magnitude, which must also be considered. For three-phase systems, the removal of triple harmonics is not required since such distortions are automatically removed from the line to line voltage.

OPTIMIZATION TECHNIQUES

Optimization is the method in which the optimal outcome is obtained under any different situations. It is commonly used in science and technology for various applications to optimize the targeted result. The structuring of the optimization challenge is to start with the identification of the design variables that are used during the optimization method. The next challenge is to find limitations linked to the optimization problem that represent a functional correlation between design variables and other design parameters.

Constraints are of two types such as equality and inequality. Equality constraints are usually more difficult to handle and need to be avoided. In the formulation method the third step is to identify the objective function as design parameters and other problem variables. The objective function can be of two types, either maximization or minimization. In this work the minimization objective function is used since the main objective of the SHE method is to minimize the harmonics. Equation (4.23) shows the objective function(O) of the SHE method in harmonics minimization for the 9-level MMI.

$$\text{Minimize}(O) = F(\theta_1, \theta_2, \theta_3, \theta_4) \quad (4.23)$$

Subjected to $0 \leq \theta_1 < \theta_2 < \theta_3 < \theta_4 \leq \frac{\pi}{2}$

12342

Based on the duality principle, the algorithm developed for the minimization problem is implemented to solve the maximization problem by simply multiplying the objective function by 1 and vice versa. The final approach is to determine the minimum and maximum limits for each design element, but this detail does not include other optimization algorithms.

Problem Formulation

For a 9-level MMI, the main purpose is to find the optimum triggering angles $\{3_1, \{3_2, \{3_3, \{3_4$ which will minimize the THD. The RMS value of the waveform illustrated in Figure 4.1, is achieved by means of

Fourier analysis as presented in the Equation (4.24).

$$\overline{2\sqrt{2}}*V$$

$$* ((\cos (n\blacklozenge) + \cos (n\blacklozenge) + \cos (n\blacklozenge) + \cos (n\blacklozenge)), \text{odd}$$
$$V_o =$$

nTI

PV(1-4) 1 2
   0, even

3 4

$$(4.24)$$

In view of the quarter wave symmetry, Equations (4.25) and (4.26) give the highest possible value of the fundamental component, when all the

triggering angles are equal to zero. i.e. $\blacklozenge_1, \blacklozenge_2, \blacklozenge_3, \blacklozenge_4 = 0$

$$\_V_{1maximum} =$$

$$\frac{2\sqrt{2}}{\_\_\_}TI * V_{PV(1-4)} * 4 \quad (4.25)$$

$$\_V_{1maximum} =$$

$$\frac{8\sqrt{2}}{\_\_\_}TI * V_{PV(1-4)} \quad (4.26)$$

For representing in per unit form,

$$V_1 \quad 2\sqrt{2} * V_{PV(1-4)} \quad TI$$

$$V\_\_\_\_\_\_\_\_V_{1(p.u)} = =$$

$$1maximum$$

$$*$$

$$\_\_\_\_\_\_\_\_\_\_\_\_\_\_\_\_\_\_\_\_\_\_\_\_nTI \quad 8\sqrt{2} * V_{PV(1-4)}$$

$V_{1(p.u)}$

$\underline{V_1}$

$=$

$V_{1maximum}$

1

$\_= 4\,(\cos(\theta_1) + \cos(\theta_2) + \cos(\theta_3) + \cos(\theta_4))\quad (4.27)$

π

$2\,2$

$V_{rms} = \int V_2\,dwt$

$$(4.28)$$

$\pi\,0$

$V = V$

$*2$

$-$

π

$$(4.29)$$

$-$

rms

$PV(1\text{-}4)$

$\pi\,(\theta_2 - \theta_1) + 4(\theta_3 - \theta_2) + 9(2 - \theta_4)$

For representing in per unit form, the Equation (4.29) is divided with (4.26), the final RMS voltage is presented in Equation (4.30)

$$V_{rms(p.u)}$$

$$\underline{V_{rms}}$$

$$=$$

$$V_{1maximum}$$

$$\underline{\qquad\qquad\qquad} TI\ TI$$

$$\underline{\quad} V_{rms(p.u)} = V_{PV(1-4)} * J64\ (\diamond_2 - \diamond_1) + 4(\diamond_3 - \diamond_2) + 9(2 - \diamond_4)$$

(4.30)

THD is described as the ratio of RMS value of all harmonic components presented in Equation (4.27) to the RMS value of the fundamental component presented in Equation (4.30). The THD value is presented in Equation (4.31).

$$\overline{\qquad\qquad}$$

$$JI,_{oo}\ V_2$$

$$THD$$

$$\text{min}$$

$$= \underline{n=2\ n}$$

$$V_1$$

$$THD$$

$$\underline{TI}$$

$$\rule{6cm}{0.4pt}$$
$$\overline{\rule{8cm}{0.4pt}} = TI * (\theta_2 - \theta_1) + 4(\theta_3 - \theta_2) + 9(2 - \theta_4)$$

$-1 \; (4.31)$

$$\min$$

$$4\cos(\theta_1) + \cos(\theta_2) + \cos(\theta_3) + \cos(\theta_4)$$

For the suggested 9-level MMI, the objective function (Min O) and constraints are specified in Equations (4.32) and (4.33).

Min O = {W(k) * (per unit equivalent - Equation (4.27)) + Equation (4.31)} (4.32)

Constraints 0 -5 θ < θ < θ < θ -5 TI

–

$$(4.33)$$
1 2 3 4 2

Where θ_1, θ_2, θ_3, θ_4 are the switching angles, W(k) is the

weighting factor which is used for making the error small enough and per unit equivalent of the output peak voltage whose value is between 0 to 1. In the

objective function (Min O), the first element (per unit equivalent- Equation

(4.27)) is the absolute value of error in adjusting the fundamental component.

BEE COLONY OPTIMIZATION

The bee system was introduced by Pham in 2007 as a derivative free optimization technique. This concept is derived by the behavior of honey bees. Bees are considered in nature as insect species with well-organized colonies. Researchers have used their behavior patterns, such as hunting and gathering, mating and location of nests to solve numerous complicated combination optimizations and problems with functional optimization.

Natural World of Bees

Honey bees in a particular colony can continuously fly in several paths to harness a great number of food sources. By practice, more bees will access the floral areas with massive quantities of nectar and pollen that can be gathered with less energy, whereas sections with fewer nectar or pollen will attract fewer bees.

There are three types of bees within colony namely skilled bees, onlooker bees, and scout bees. Within a specific food supply, skilled bees bear details on the location and quantity of nectar. They do a dancing activity to pass the knowledge to onlooker bees. The dancing time represents the quantity of nectar in a food source. This dancing activity involves three bits of details on a floral patch: its range from the hive, its position and its fitness (or quality). This dancing activity is needed for the interaction between colonies and the knowledge helps the colonies to send their bees accurately to floral areas, with no guides or maps.

The knowledge provided from dancing helps the colony to determine the relative importance of the numerous locations in conjunction with both the quality of the food that is received and the resources necessary to produce it. The onlooker selects the source of food depending on the quantity of nectar in a food source. A good source of food tends to attract more onlooker bees.

Scout bees search for habitat to discover different sources of food. Scout bees monitor the exploration process, while skilled bees and onlookers play an exploitative role. The sources of food are taken as feasible solutions to a constraint in this method. The source of food is a D-dimensional matrix, where D is the number of factors for optimization. The quantity of nectar in a food supply decides the fitness value as presented by Falehi (2019).

Computation of Switching Angles

By implementing its bees to effective fields, a colony thrives. The BCO is used in the suggested method to determine the optimum switching angles needed by a solar fed MMI. The cost function for the 9-level MMI is presented in Equation (4.34)

$$V V 1 V^2 1 V^2$$

$$V\text{\_\_\_\_\_\_\_\_\_}F 1^1{}_{13}$$

$$1 3 V_1 V_1$$

$$1 V_{72}$$

$$\overline{}$$

$$7 V_1$$

(4.34)

Subjected to

$$::; P$$

$$TI$$

$$< P < P < P ::;$$

$$-$$

$$1 2 3 2$$

[1]where V^* is the required fundamental harmonic, V_1 is calculated using Equation (4.14).

The fitness function for the 9-level MMI is presented in Equation (4.35)

Fitness Function(F)

___

$$(4.35)$$

$$_F F\ 1$$

Where,F is the cost function calculated using Equation 4.34.

When the fundamental harmonic component exceeds the set point by 1% the first term in the Equation (4.34) finely tunes the harmonic component to the power of 4. Due to the use of a power of 4, the corresponding penalties are negligible for deviations below 1 %. The second, third and fourth term refers to the elimination of 3rd,5th and 7th order harmonics. These lower order harmonic terms in Equation (4.34) neglects harmonics below 2% of the fundamental. However, when any harmonic

exceeds this limit, the objective function is subject to a power penalty of 2.

Lastly, each harmonic ratio is determined by inverse of its harmonic order. By this method the lower order harmonics are eliminated from the fundamental component.

The algorithm of bees includes the following steps:

Step 1: Random loading of 40 preliminary food sources. The number of preliminary sources of food is half the colony of bees.

Step 2: Skilled bees are dispatched to food sources to measure the quantity of nectar and their fitness. A single skilled bee is implemented for each food source. The number of sources of food is the same as the number of skilled bees. The skilled bees adjust the solutions saved in their memory through a search in the vicinity of their food source. The skilled bees save the new solution when its fitness is greater than the older one. Skilled bees return to the hive and communicate the ideas with onlooker bees.

Step 3: On-looker bees picks the appropriate food sources using a probability method. More quantity of nectar food sources lure more on-looker bees. On-looker bees are directed to the chosen food sources. On-looker bees progress the selected solutions and analyze their fitness. Similar to skilled bees, the on-looker bees also save a renewed solution if its fitness is greater than the older solution.

Step 4: The sources of food which are not upgraded after numerous iterations are discarded. So, the skilled bee is directed to find new sources of food as a scout bee. The discarded food source is substituted by the new food source.

Step 5: The best source of food is saved. At the conclusion of the 80

iteration, the maximum number of iterates is fixed as a termination criterion. If not completed, for the next iteration, the method goes to step 2.

GENETIC ALGORITHM

Genetic algorithm (GA) is a derivative free probabilistic optimization algorithm used to tackle both restricted and unrestricted problems of optimization, based on the biological evolution. It deals with the maximization and minimization issues. A population of definite solutions are continuously changed by the genetic algorithm. In each step, GA selects individuals to be parents from the current population and uses them for the next generation to produce offspring. The population is "evolving" toward an optimal solution over successive generations.

GA was initially developed by John Holland, the University of Michigan in 1975. GAs transform each element into a binary bit string called chromosomes in a solution space and every point has a health value that is typically equivalent to the measured objective function for maximization. GAs generally hold a range of points as a population (or gene pool), which is then obtained repeatedly towards a best fitness value. The GA creates a new population for every generation that uses genetic operators, such as crossover and mutation. Members with higher levels of fitness are more likely to succeed and engage in combination operations. The population consist of individuals with good fitness values after several generations.

Computation of Switching Angles

GA is a search algorithm that resembles the biological evolution process. In the suggested method, GA is used for determining the optimal switching angles for a 9-level SMMI. GA consists of 6 major stages. They are: 1) Creation of preliminary chromosomes, 2) Population, 3) Fitness evaluation, 4) Crossover, 5) Mutation, and 6) Conclusion.

- Creation of preliminary chromosomes:

The primary procedure of GA is the formation of primary chromosome. The preliminary chromosomes are formed from the lowest to the highest values of the total chromosome. A 9-level MMI require four switching angles, thus each chromosome for this application will have four switching angles. The next step is to calculate the fitness function after the production of preliminary chromosomes.

- Population

The GA begins with a set of chromosomes known as the population. The size of the population is an essential variable that needs to be tuned. The complexity of the problem, which is reflected by the size of the search space, is a factor to be considered while fixing the size of the population. The preliminary population is randomly produced. The higher population might increase the rate of convergence but it also increases the execution time. The population considered has 20 chromosomes, each containing four triggering angles for a 9-level MMI. The population is loaded with random angles between 0_0 and 90_0 considering the quarter-wave symmetry of output.

- Fitness function

Fitness is an essential approach in GA which is used to determine the best chromosome. Fitness functions are objective functions that are used to evaluate a particular solution represented as chromosomes in a population. There are two types of fitness functions: the first one, the fixed type does not allow the fitness function to change. In the second case, the fitness function is mutable. In this work the mutable fitness function is chosen because the fitness function changes different levels of output.The most important factor in deciding a fitness function is that it should correlate to the problem closely and is simple enough to be computed quickly.

The cost function is the most essential factor for the GA in determining the fitness or vitality of every chromosome. The goal of the suggested method is to reduce different harmonics; therefore, the cost function must be associated to such harmonics. In this work, the 3_{rd}, 5_{th}, and 7_{th} harmonic orders at the output of a 9-level MMI are to be minimized. Then the cost function (f) can be selected as the sum of these harmonics normalized to the fundamental given in Equation (4.36) .

$$f(\alpha_1, \alpha_2, \alpha_3, \alpha_4) = 100 *$$

$$IV_3I + IV_5I + IV_7I$$

$$\frac{1}{\rule{3cm}{0.4pt}} IV I \quad (4.36)$$

For every chromosome, a multilevel output is produced by utilizing the triggering angles in the chromosome and the essential harmonic magnitudes are computed using FFT methods. The triggering angle set delivering the maximum fitness value (minimum THD) is the finest solution of the first iteration.

• Crossover operation

Crossover procedure among two chromosomes is carried out in order to generate a new chromosome. The crossover process is carried out on the basis of the crossover rate. The crossover is used to pick genes and produce a new child chromosome. The fitness value is then provided to this new child chromosome after creation of a new chromosome. The different types of crossover used in GA are single crossover , dual crossover , multiple crossover , fixed crossover and adaptable crossover. The next approach after the crossover is mutation.

- Mutation operation

The mutation operator imparts a small change at a random position within a chromosome. The mutation procedure is carried out depending on the normal weak mutation rate. The mutation is achieved by random mutation of the genes depending on the specified mutation rate.

- Termination

In the termination phase, the finest chromosome is preferred depending on the fitness function. The above procedure is repeated till the

maximum number of 100 iterations. At the end of the termination method, the

best switching angles are obtained. The coding for GA is written in M file and executed to obtain proper switching angles for minimum THD

PARTICLE SWARM OPTIMIZATION

Particle Swarm Optimization (PSO) has been developed in 1995 by Dr James Kennedy and Dr Russell Eberhart as a population-based, probabilistic optimization algorithm. PSO is motivated by social activity of bird flocking or fish schooling. The PSO algorithm maintains a swarm of individuals (called particles), where each particle represents a candidate solution. In PSO, the potential solutions, called particles, fly through the problem space by following the current optimum particles. Particles follow a simple behavior: emulate the success of neighboring particles and their own achieved successes.

The implementation of PSO involves the following functional variables. An 'i' particle in the group has three features, such as:

a. Location in search space X_i,

a. Velocity Y_i

b. Personal best $P_{best(i)}$

Another significant feature of the swarm is the global best location G_{best} which is the best location of the particles in the total population. The location and velocity are reorganized by Equation (4.37) and (4.38)

respectively.

$$y_{ij}(K+1) = w(K)y_{ij}(K) + S_1R_1(P_{best(ij)}(K) - X_{ij}(K)) + S_2R_2(G_{best(i)}(K)$$

$$- X_{ij}(K)) \quad (4.37)$$

$$X_{ij}(K+1) = X_{ij}(K) + Y_{ij}(K+1) \quad (4.38)$$

Where X_{ij} is the position of particle 'i' in dimension 'j' , Y_{ij} is the velocity of the particle 'i' in dimension 'j' , the number of iteration is given by K , $w(K)$ is the weight function, S_1, S_2 are the acceleration coefficients, R_1, R_2

are the random numbers which are uniformly distributed and limits between $[0,1]$.

Equation (4.37) is composed of three attributes such as inertial component, the cognitive component, and social component. The inertial behavior of the bird to fly in the previous path is known as the inertial component. The cognitive component includes the memory of the bird about its previous position whereas the social component denotes the memory of the bird about the best position among the particles and collaborative behavior of the particles to find the global optimal solution. Equation (4.38) specifies the calculation of new velocity by the summation of the previous position and new velocity.

Computation of Switching Angles

The algorithmic steps involved in solving the SHE problem for the 9-level MMI is as follows:

Step 1: Start with random position (X_i) and velocity (V_i) vectors. Each particle in the population is randomly initialized(switching angles) between 0° and 90°. Likewise, the velocity vector of every particle

is to be randomly produced between $-V_{max}$ and V_{max}.

Step 2: Calculate the particles by using the fitness function of the harmonic reduction problem. The triggering angles (1 to 4) can be calculated such that the 3_{rd} 5_{th} and 7_{th} order of harmonics can be removed from the output of the MMI to minimize the cost function. The fitness function is given in Equation (4.39) and is derived based on the expression used for an 11-level inverter by Taghizadeh & Hagh (2010).

$$f(\theta_1, \theta_2, \theta_3, \theta_4) = 100 * \left[IM_i - \frac{IV_iI}{IV_{PV(1-4)}I} \right]I +$$

$$IV_3I + IV_5I + IV_7I \; IV_{DC(1-4)}I$$

$$I \quad (4.39)$$

Where, M_i is the modulation index. On the evaluation of fitness function, P_{best} and G_{best} can be obtained.

Step 3: Update the P_{best} (P_i) position of the particles. If the present location of the ith particle is better than its prior best location, substitute P_i with the present location X_i. Moreover, if the best location of the

individual bests of the particles is greater than the location of the global best, substitute G_{best} (P_g) with the best location of the individual bests.

Step 4: Rearrange the velocity and location vectors. All the particles in the population are rearranged by velocity and location update rules given in Equations (4.37) and (4.38).

Step 5: The maximum number of 100 iteration count is set as termination

criteria. If the iteration counter reaches maximum, stop; else, increase the iteration counter and go back to step 2.

SIMULATION RESULTS

This section explains the simulation and experimental results of the MMI using the optimization techniques.

Implementation of SHE for 9-Level MMI

Selective Harmonic Elimination Technique decomposes the 9-level output waveform using Fourier series. A group of nonlinear equations are solved to obtain the triggering angles which are instantly used to trigger the switches of MMI to produce a harmonic less stepped waveform. In this chapter the complexity in solving the nonlinear equations are made easy by using optimization techniques. Figure 4.2 shows the switching pulses of the unidirectional, bidirectional and the P1,P2 switches of 9-level MMI.

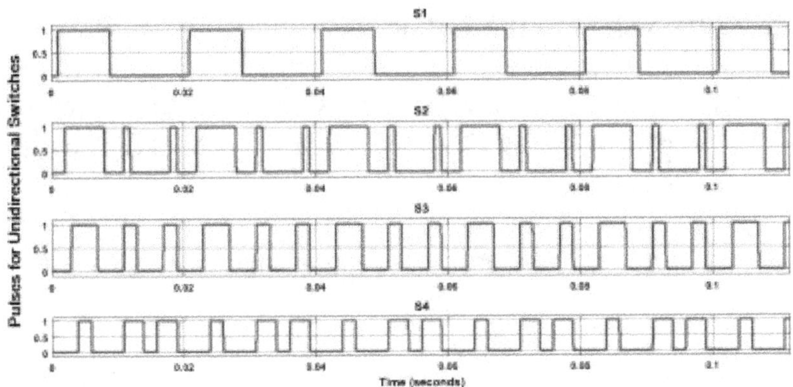

(a)

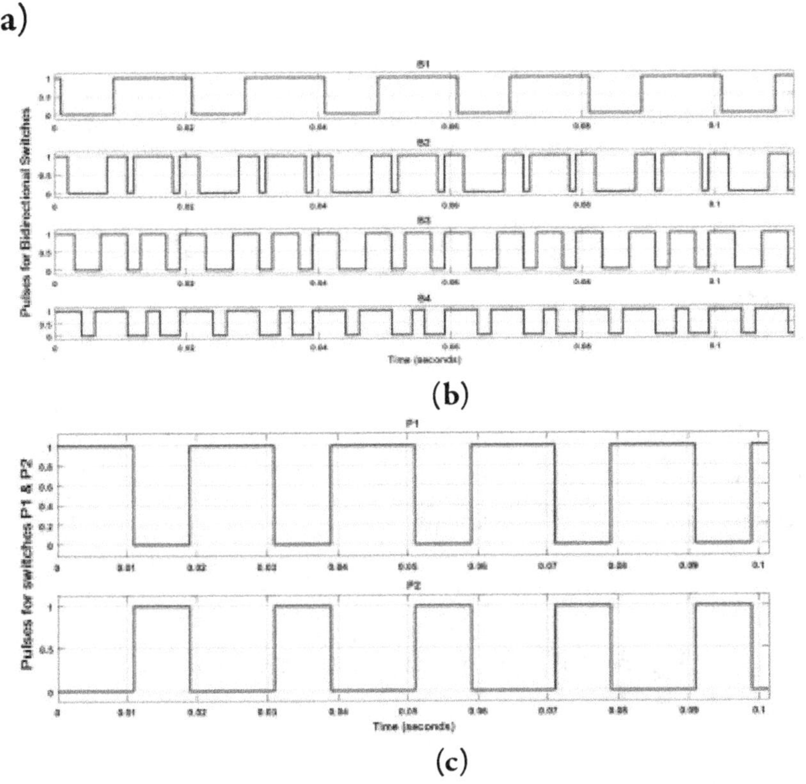

(b)

(c)

Figure 4.2 Switching Pulses for (a) unidirectional (b) Bidirectional

(c) P1 and P2 Switches of 9-Level MMI

4.7.1.1 Bee colony optimization

Table 4.1 presents the variables of BCO used for the execution of the program. The BCO program is run in Matlab and the triggering angles are

obtained. Table 4.2 presents the triggering angles for different M_i. From

Table 4.2, it is inferred that when the modulation index M_i is 0.9, the optimum switching angles give minimum THD of 10.43%.

Table 4.1 BCO Parameters

Parameter	Value
Number of skilled bees	20
Number of scout bees	20
Number of sites selected for neighbourhood search	15
Number of elite sites	4
Stopping criteria	80 iterations

Table 4.2 Triggering Angles for different Modulation Index and THD for BCO

Modulation Index	Switching Angles (Degrees)				o/o THD
	1	2	3	4	
0.6	24.21	31.28	49.88	71.56	15.14
0.7	28.36	36.54	47.19	77.43	13.86
0.8	31.89	39.15	46.25	64.57	12.52
0.9	13.41	29.88	42.79	66.39	10.43

Figure 4.3 presents the fitness values of different generations. It is observed from Figure 4.3 that, for various modulation indexes ranging from

0.4 to 1.2, the value of cost function in Y axis is between 10^{-12} and 10^{-14};

hence, the efficiency of this method for solving SHE problem is evident. Figure 4.4 presents the output and FFT analysis of the 9-level MMI using BCO. From the FFT analysis, it is concluded that the THD obtained using BCO is 10.43%.

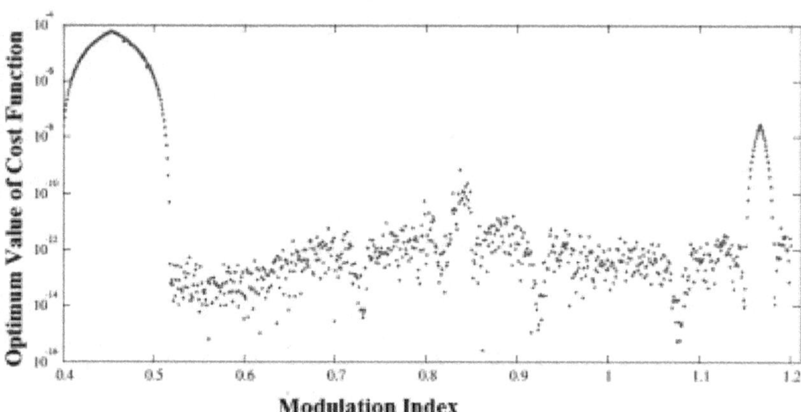

Figure 4.3 Modulation Indices Vs Optimum Value of Cost Function Using BCO

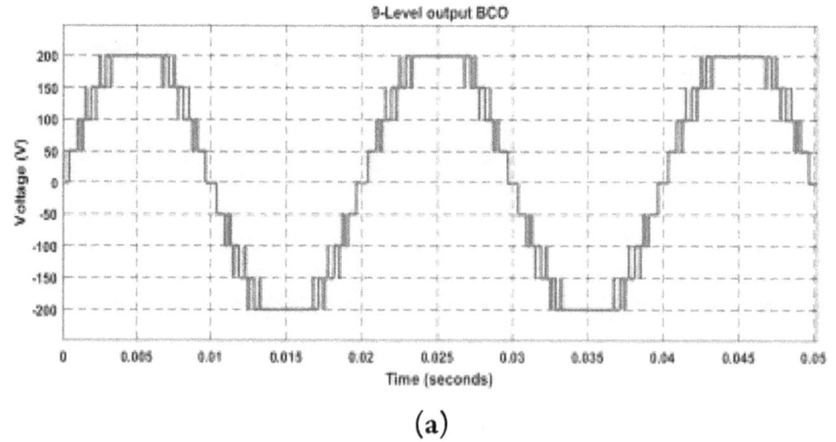

(a)

Figure 4.4 (Continued)

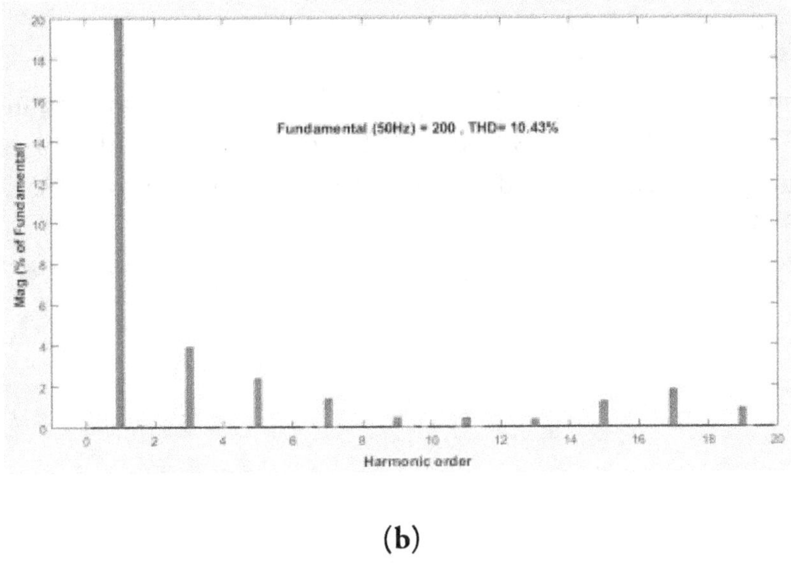

(b)

Figure 4.4 (a) 9-Level output voltage (b)FFT analysis of 9-Level MMI Using BCO

4.7.1.2 Genetic algorithm

For the proposed 9-level MMI, Table 4.3 shows the GA parameters. Table 4.4 shows the switching angles required to reduce the 3rd,5th and 7th order harmonics for several ranges of modulation indices using GA. Figure 4.5. presents the fitness values for different modulation index. It is

observed from Figure 4.5 that, the fitness values fall in the range of 10^{-12} and

10^{-14} for different modulation indexes and this value gradually decreases

from 0.9 to 1.2 of the modulation index value. Therefore, the best fitness values are found in the region where the modulation index is 0.9. Table 4.2, it is inferred that when the M_i is 0.9, the optimum switching angles give

minimum THD of 7.11%. Figure 4.6 presents the output and FFT analysis of the 9-level MMI using GA. From the FFT analysis, it is concluded that the THD obtained using GA is 7.11%.

Table 4.3 GA Parameters

Parameter	Value
Population Size	24
Crossover Rate	0.5
Mutation Rate	0.234
Stopping Criteria	100 iterations

Table 4.4 Triggering Angles for different Modulation Index and THD using GA

Modulation Index	Switching (Degrees)	Angles			o/o THD
	1	2	3	4	
0.6	26.92	36.21	49.28	63.56	13.44
0.7	23.73	39.14	48.39	60.18	11.07
0.8	19.36	30.05	45.77	68.35	9.81
0.9	12.34	28.63	39.12	57.53	7.11

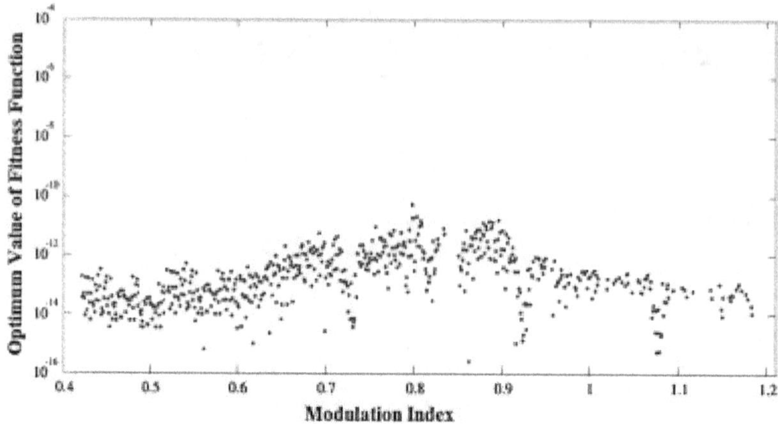

Figure 4.5 Modulation Indices Vs Optimum Value of Fitness Function Using GA

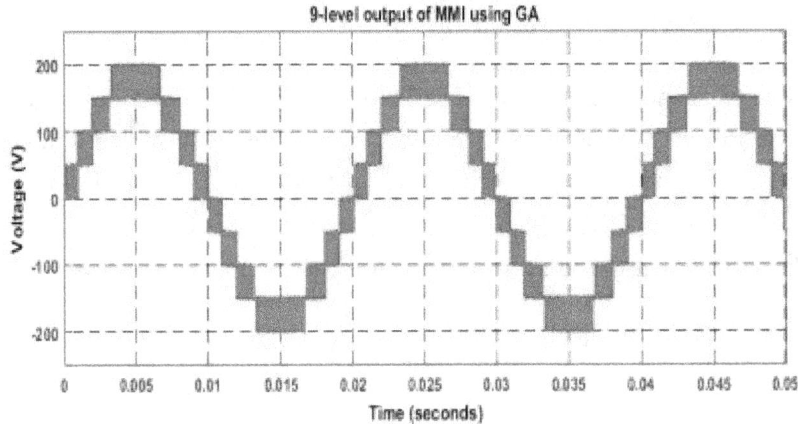

(a)

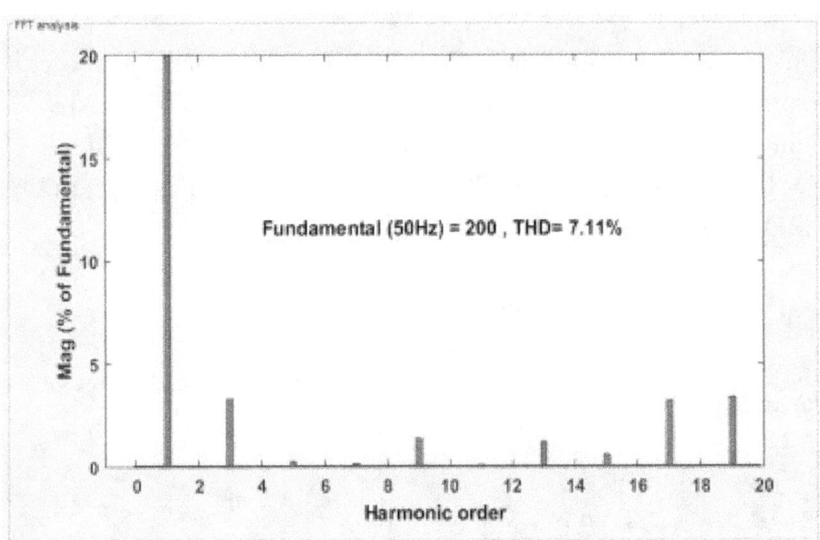

(b)

Figure 4.6 (a) 9-Level output voltage (b)FFT analysis of 9-Level MMI Using GA

4.7.1.3 Particle swarm optimization

Table 4.5 shows the PSO parameters used for the execution of the program. Table 4.6 shows the switching angles required to reduce the 3rd, 5th and 7th order harmonics for several ranges of modulation indices using PSO. Figure 4.7 shows the relation between the fitness value and the modulation

index. From Figure 4.7 it is observed that the fitness value lies between 10^{-12}

and 10^{-14} for the modulation index 0.1 to 0.9. It is also noted that the fitness

value has a steep raise from 0.7 to 0.8 of modulation index. The best fitness values are obtained at the modulation index 0.9. Figure 4.8 presents the output and FFT analysis of the 9-level MMI using PSO. From the FFT analysis, it is concluded that the THD obtained using PSO is 5.15%.

Table 4.5 PSO Parameters

Parameters	Value
S1, S2	0.3
R1, R2	0.9 [0 to 1]
w(k)	0.9
Stopping criteria	100 iterations

Table 4.6 Triggering Angles for different Modulation Index and THD for PSO

Modulation Index	Switching Angles (Degrees)				%THD
	1	2	3	4	
0.6	13.42	39.23	55.11	70.21	8.34
0.7	12.65	35.75	58.94	72.62	7.51
0.8	11.91	33.53	60.46	76.83	6.12
0.9	10.36	31.14	64.92	82.72	5.15

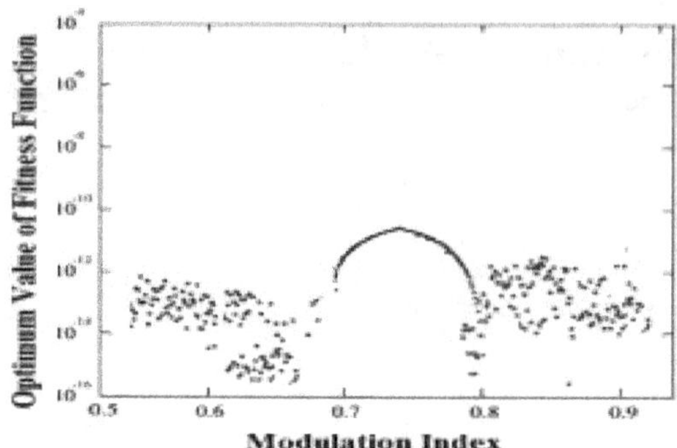

Figure 4.7 Modulation Indices Vs Optimum Value of Fitness Function Using PSO

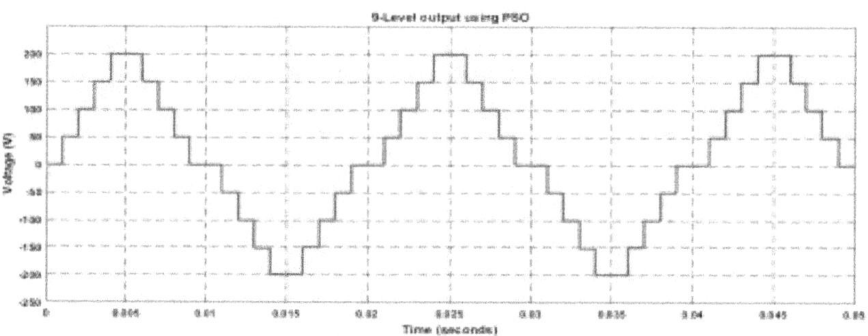

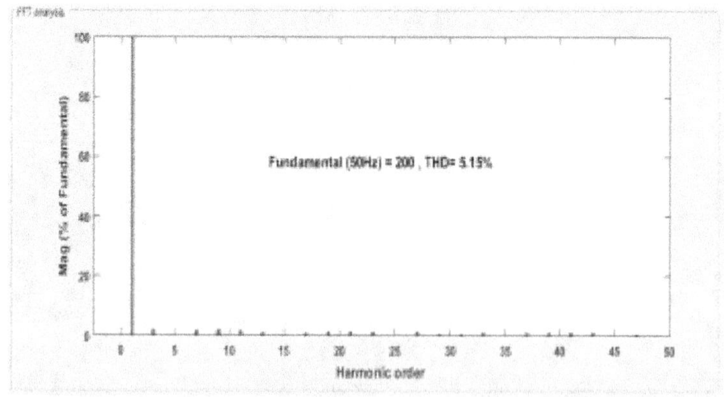

(a)

(b)

Figure 4.8 (a) 9-Level output voltage (b)FFT analysis of 9-Level MMI Using PSO

The output of 9-level MMI consist of H_3, H_5, H_7 order of harmonics. These harmonic contents have to be minimized since they are dominant and may cause damage to the system. Table 4.7 presents the THD values of the 9-level MMI for various modulation index values using BCO, GA and PSO optimization techniques. From Table 4.7 it is evident that the total THD % as well as the harmonic contents of the proposed 9-level MMI are very less for the PSO algorithm when compared with the other two techniques.

Table 4.7 Performance Comparison of THD and Harmonic components of 9-level MMI with BCO,GA and PSO methods

Method	Modulation Index	Switching Angles (Degrees)				%THD	H$_3$ (%)	H$_5$ (%)
		1	2	3	4			
BCO		24.21	31.28	49.88	71.56	15.14	1.08	0.68
GA	0.6	26.92	36.21	49.28	63.56	13.44	0.95	0.28
PSO		13.42	39.23	55.11	70.21	8.34	0.49	0.45
BCO		28.36	36.54	47.19	77.43	13.86	0.78	0.31
GA	0.7	23.73	39.14	48.39	60.18	11.07	0.57	0.06
PSO		12.65	35.75	58.94	72.62	7.51	0.39	0.46
BCO		31.89	39.15	46.25	64.57	12.52	0.68	0.26
GA	0.8	19.36	30.05	45.77	68.35	9.81	0.45	0.43
PSO		11.91	33.53	60.46	76.83	6.12	0.16	0.25
BCO		13.41	29.88	42.79	66.39	10.43	0.52	0.04
GA	0.9	12.34	28.63	39.12	57.53	7.11	0.37	0.42

PSO   10.36 31.14 64.92 82.72   5.15   0.05 0.01 0

H_3, H_5, H_7-Third, Fifth, Seventh order harmonics

338

Implementation of SHE for 27-level MMI

Figure 3.2a shows the basic structure of MMI with four PV sources. If the input PV sources are in the ratio of 1:2:3:7 then a 27-level MMI is

obtained. In selective harmonic elimination method, N switching angles are

required to produce 2N + 1 level of output voltage with N DC sources. To reduce N - 1 order of harmonics in the output voltage,N switching angles are obtained by solving N nonlinear equations. Therefore, for a 27-level MMI,13

switching angles are required to eliminate the 3,5,7,9,11,13,15,17,19,21, 23, and 25 order of harmonics. The thirteen nonlinear equations can be set up as follows:

$$B = \frac{4V}{\pi}[\cos\theta + \cos\theta + \cdot + \cos\theta],$$

1 1TI 1 2 13

$$B = \frac{4V_{PV}}{3\pi}[\cos3\theta + \cos3\theta + \cdot + \cos3\theta],$$

3 3TI 1 2 13

$$B = \frac{4V_{PV}}{5\pi}[\cos5\theta + \cos5\theta + \cdot + \cos5\theta],$$

5 5TI 1 2 13

$$B = \frac{4V_{PV}}{7\pi}[\cos7\theta + \cos7\theta + \cdot + \cos7\theta],$$

7 7TI 1 2 13

$$B = \frac{4V_{PV}}{9\pi}[\cos9\theta + \cos9\theta + .. +\cos9\theta]$$

9 9TI 1 2 13

$$B = \frac{4V_{PV}}{11\pi}[\cos11\theta + \cos11\theta + .. +\cos11\theta],$$

11 11TI 1 2 13

$$B = \frac{4V_{PV}}{13\pi}[\cos13\theta + \cos13\theta + .. +\cos13\theta],$$

13 13TI 1 2 13

$$B = \frac{4V_{PV}}{15\pi}[\cos15\theta + \cos15\theta + .. +\cos15\theta],$$

15 15TI 1 2 13

$$B_{17} = \frac{4V_{PV}}{17\pi}[\cos 17\theta_1 + \cos 17\theta_2 + .. + \cos 17\theta_{13}],$$

$$B_{19} = \frac{4V_{PV}}{19\pi}[\cos 19\theta_1 + \cos 19\theta_2 + .. + \cos 19\theta_{13}],$$

$$B_{21} = \frac{4V_{PV}}{21\pi}[\cos 21\theta_1 + \cos 21\theta_2 + .. + \cos 21\theta_{13}],$$

$$B_{23} = \frac{4V_{PV}}{23\pi}[\cos 23\theta_1 + \cos 23\theta_2 + .. + \cos 23\theta_{13}],$$

$$B_{25} = \frac{4V_{PV}}{25\pi}[\cos 25\theta_1 + \cos 25\theta_2 + \cdot + \cos 25\theta_{13}]$$

$$] \tag{4.40}$$

Figure 4.9 presents the switching sequence of the unidirectional, bidirectional and P1, P2 switches of 27 level MMI. The switching sequences are generated by storing the switching angles which are obtained using the optimization techniques in the lookup table. Figure 4.10 shows the relation between the modulation index and the switching angles. Table 4.8 shows the switching angles of the 27 level MMI for one complete cycle.

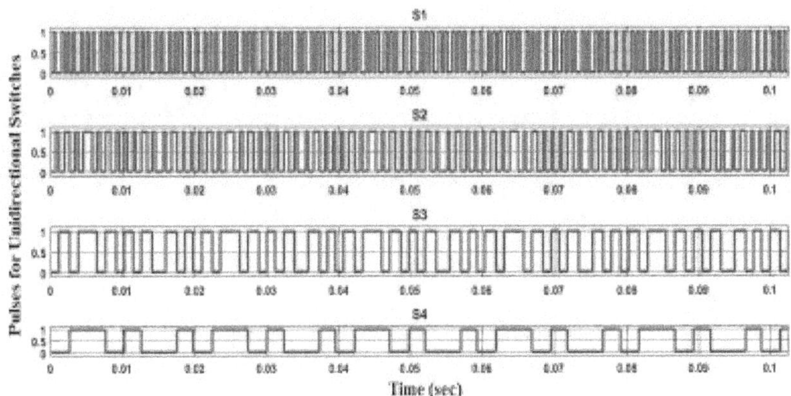

(a)

Figure 4.9 (Continued)

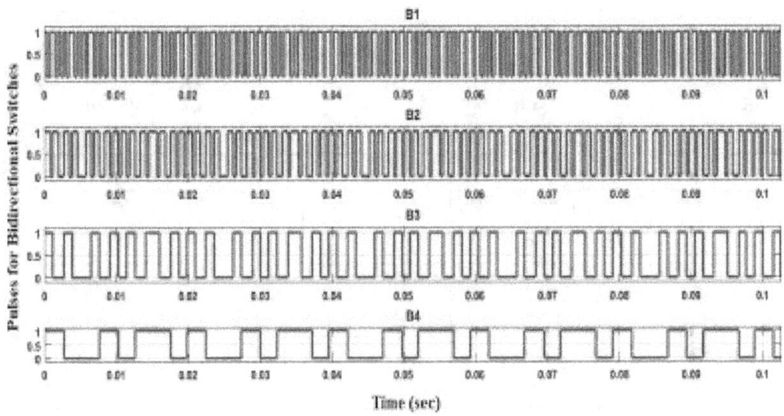

(b)

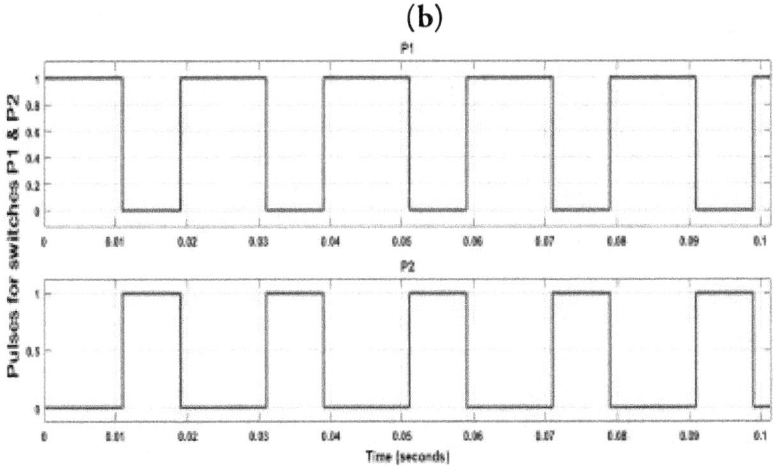

(c)

Figure 4.9 Switching Pulses for (a) unidirectional (b) Bidirectional (c) P1 and P2 Switches of 27-Level MMI

$M_i = 0.9$				
Switching angle	1st Quarter	2nd Quarter	3rd Quarter	4th Quarter
β_1	8.9245	157.370	171.890	350.716
β_2	14.846	160.384	182.413	345.46
β_3	21.769	157.131	189.321	338.204
β_4	26.693	147.878	190.619	323.954
β_5	36.625	143.625	197.172	321.692
β_6	40.53	134.	203.4	311.4

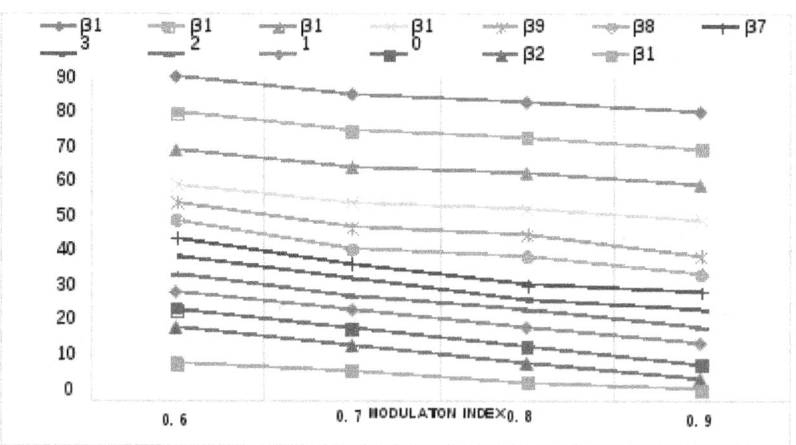

Figure 4.10 Modulation Index Vs Switching Angles for 27-Level MMI Table 4.8 Switching Angles for 27-level inverter

For the 27-level Inverter the twelve nonlinear Equations are solved using BCO,GA, and PSO algorithms. For the implementation of BCO to the 27-level MMI, the fitness function is given in Equation (4.41).

$$VF = 100 * \left(\frac{V^* - V_1}{V_1} \right)^2$$

$$+ \frac{1}{(3,5,7\ldots25)}$$

$$(50 *$$

$$3,5,7,..25)$$

$$\frac{}{V_1}$$

$$(4.41)$$

The objective functions required to calculate the switching angles for 27-level MMI using GA and PSO are given in Equation (4.42) and (4.43).

$$f(\theta_1, \theta_2, \theta_3, \theta_4, \theta_5, \theta_6, \theta_7, \theta_8, \theta_9, \theta_{10}, \theta_{11}, \theta_{12}, \theta_{13})$$

$$|V_3| + |V_5| + |V_7| +$$

$$|V_9| + |V_{11}| + |V_{13}|$$
$$+|V_{15}| + |V_{17}| + |V_{19}|$$

$$= 100 . \frac{+|V_{21}| + |V_{23}| + |V_{25}|}{|V_1|}$$

344

(4.42)

$f(\Diamond_1, \Diamond_2, \Diamond_3, \Diamond_4, \Diamond_5, \Diamond_6, \Diamond_7, \Diamond_8, \Diamond_9, \Diamond_{10}, \Diamond_{11}, \Diamond_{12}, \Diamond_{13})$

r I I

I

__________ = IIM_i - IV

IV_1I

I + 100
I
I PV(1-4)
I I

$$IV_3I + IV_5I + IV_7I + IV_9Il$$

$+IV_{11}I + IV_{13}I + IV_{15}I\ I$
$+IV_{17}I + IV_{19}I + IV_{21}I\ I$

$* \; \underline{+IV_{23}I + IV_{25}I}\ I$

(4.43)

$IV_{PV(1-4)}I\ I$
I I J

Figure 4.11 shows the best fitness values of BCO, GA and PSO algorithms. From Figure 4.11 it is observed that the PSO algorithm suits bests since it has a lower best fitness value and lower mean value. For simulation

purpose the input PV sources to the MMI is taken as $V_{pv} = 24V$, $V_{pv2} =$

48V, $V_{pv3} = 72V$, $V_{pv} = 168V$. Therefore the peak voltage obtained is 312V.

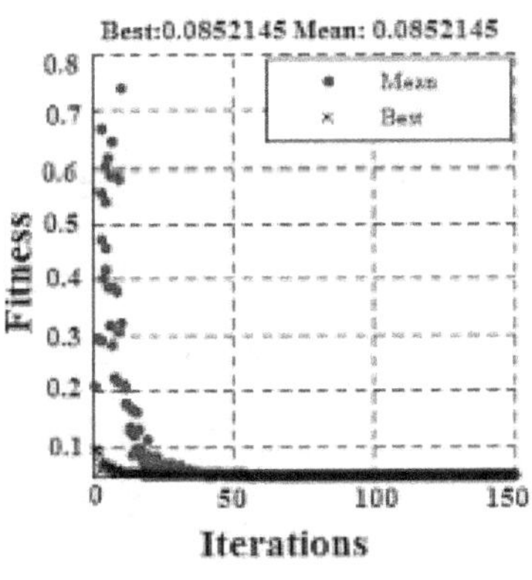

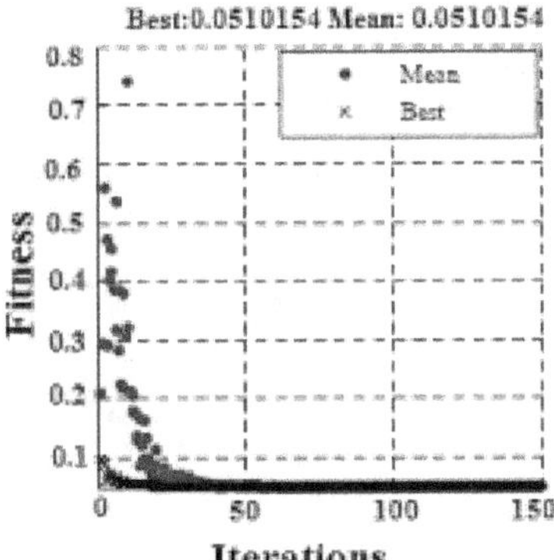

(a) (b)

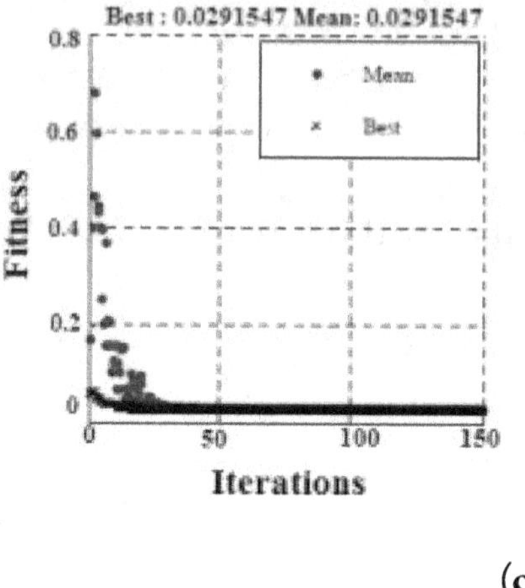

(c)

Figure 4.11 Best fitness value for a) BCO b) GA c) PSO for 27-Level MMI

Figure 4.12 shows the output waveform of 27-level MMI with its FFT analysis using the BCO algorithm. From the FFT analysis, it is concluded that the THD obtained using BCO is 7.21 %. Figure 4.13 shows the output waveform of 27-level MMI with its FFT analysis using the GA algorithm. From the FFT analysis, it is concluded that the THD obtained using GA is 6.31%. Figure 4.14 shows the output waveform of 27-level MMI with its FFT analysis using the PSO algorithm. From the FFT analysis, the PSO algorithm gives a better result than the GA and BCO. The THD obtained is 4.76% which is very less and is within the IEEE519 standards.

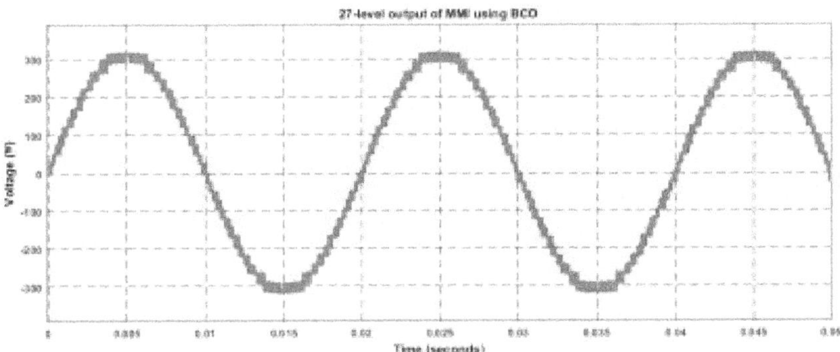

(a)

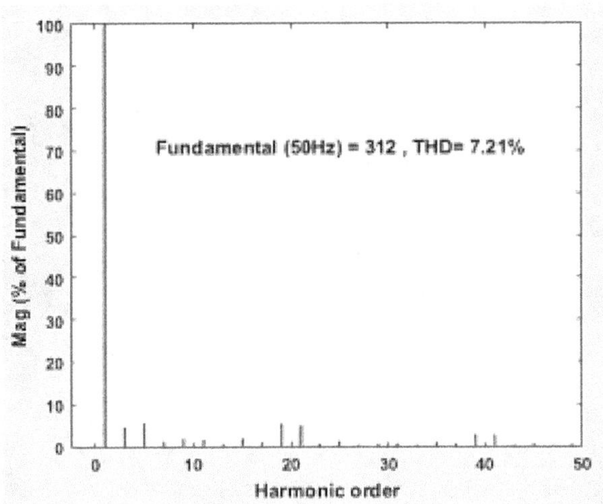

(b)

Figure 4.12 (a) 27-Level output voltage (b)FFT analysis of 27-Level MMI Using BCO

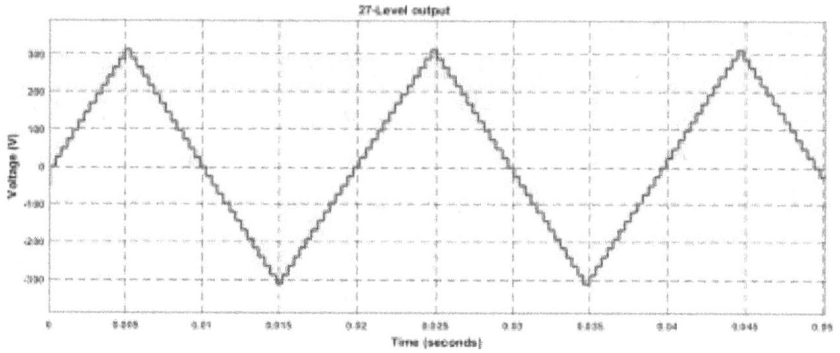

(a)

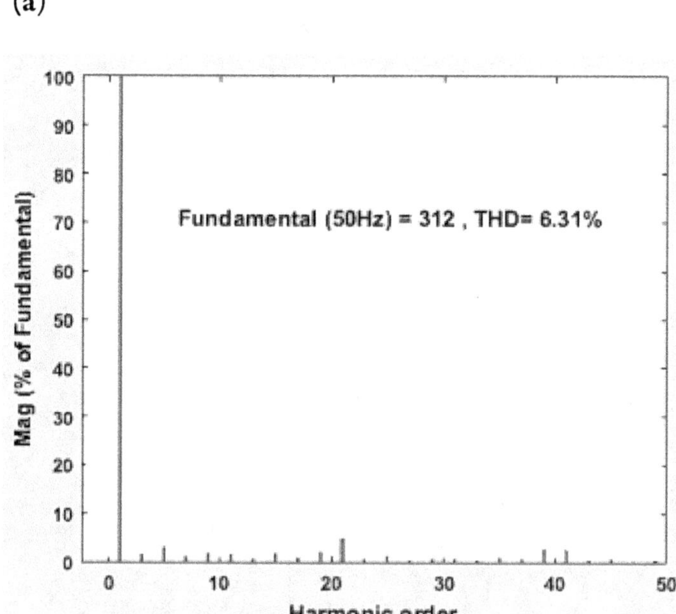

(b)

Figure 4.13 (a) 27-Level output voltage (b)FFT analysis of 27-Level MMI Using GA

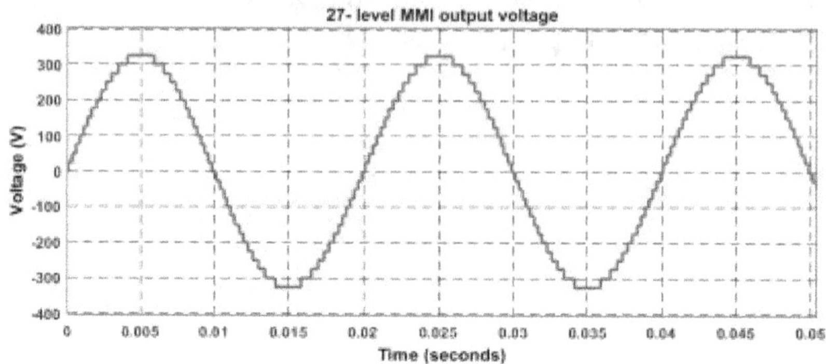

(a)

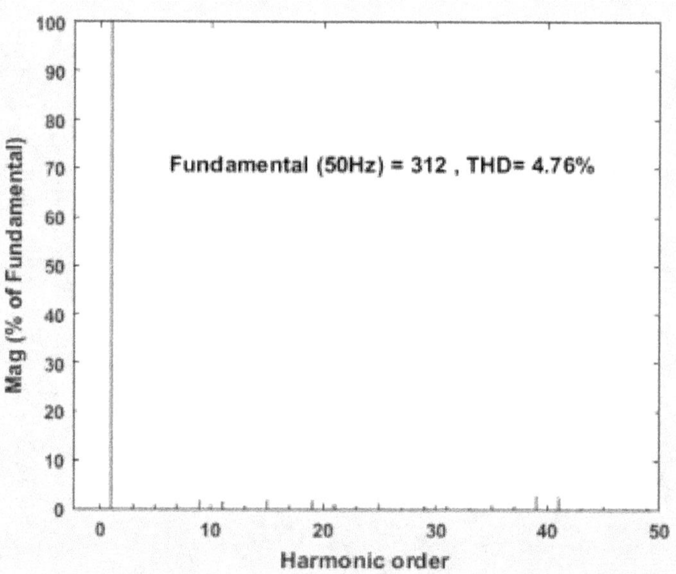

(b)

Figure 4.14 (a) 27-Level output voltage (b)FFT analysis of 27-Level MMI Using PSO

EXPERIMENTAL RESULTS

` For the hardware implementation, the input PV sources are obtained from the 1kWp PV plant. Figure 4.15 shows the setup of a 1kWp PV plant. The PSO algorithm is used in the hardware implementation for computation of switching angles. Figure 4.16 shows the hardware setup including the four PV sources, FPGA SPARTAN 6E processor, Digital Storage Oscilloscope (DSO), and Load.

In the processor, the offline procedure is utilized where the switching angles are pre-calculated and then programmed. Using the Matlab software, the PSO code is run for generation of switching angles. These switching angles which are in .m files are converted to .vdl using the HDL coder . This conversion is necessary since the FPGA kit works with .vhd code. The processor finds the relevant switching angles and transfers it to the driver circuit which is connected to IGBT for turning it ON.

Figure 4.15 Solar Plant of rating 1kWp

The desirable parameters of Xilinx SPARTAN 6E processor are:

i. Very low cost,

i. Advanced 90-nanometer manufacturing technology,

i. Dual voltage dual-standard selector pins,

i. Enhanced dual-band data rate (DDR) support,

i. Efficient compact multiplexers with wide logical capability,

i. Fast look-ahead carrying logic and vii) eight digital clock managers which make the processor suitable for SHE based applications. Figure 4.17 illustrates the experimental output of the suggested 9-level MMI with its FFT analysis. The THD obtained for the 9-level MMI is 8.44%.

Figure 4.16 Hardware set up of MMI with SHE

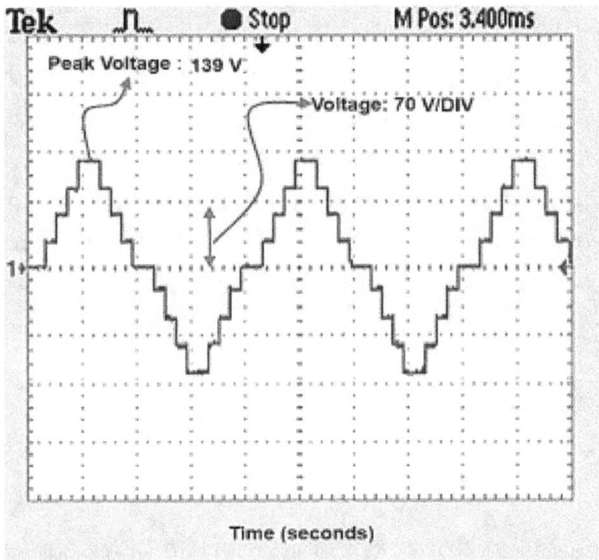

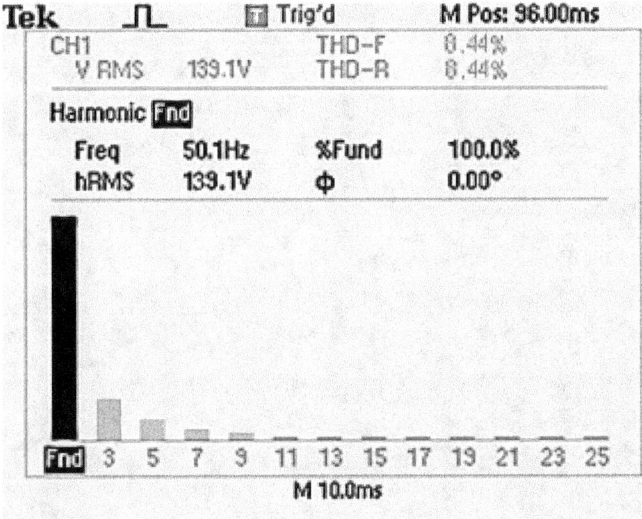

(a)

(b)

Figure 4.17 Experimental output of the 9-level MMI with its FFT analysis.

Figure 4.18 shows output waveform along with the Fast Fourier Transform (FFT) analysis of the 27-level MMI, where the THD achieved in this technique is about 6.67%. The comparison of the methods proposed for SHE is given in Table 4.9 which shows that PSO based method provides better results while comparing with the other two optimization techniques GA, and BCO. Table 4.10 presents the comparison of the SHE problem with various algorithms.

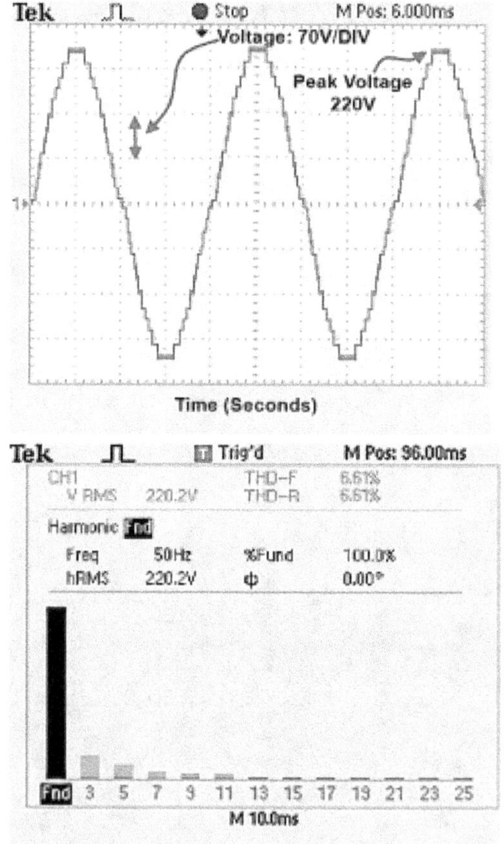

Figure 4.18 Experimental output of the 27-level MMI with its FFT analysis

Table 4.9 Comparison of Results for the Proposed Methodologies

S. No.	Method	Level	THD (%) Simulation	Experimental
1	BCO	9	10.43	-
2	GA	9	7.11	-
4	PSO	9	5.15	8.44
5	BCO	27	7.21	-
7	GA	27	6.31	-
8	PSO	27	4.76	6.61

Table 4.10 Comparison with Other Modulation Methods

S.No.	Authors	Levels	Methodology	THD
1	Panda *et al.* (2019)	7	SHE-PSO	19.89%
			SHE-FSO	14.66%
			SHE-PSFSO	10.47%
2	Haghdar (2019)	15	TLBO	1.45%
3	Kar *et al.* (2019)	11	WOA	7.51%
4	Yong Ning *et al.* (2019)	3	GA-PSO	6.32%
5	Panda & Panda (2018)	7	PSO	11.81%
6	Memon *et al.* (2018)	7	APSO	12.52%
7	Mahato *et al.* (2019)	9	SHE-PWM	13.68%
8	Arun & Noel (2018)	9	MCPWM	5.67%
9	Samadaei *et al.* (2018b)	17	NLC-PWM	3.06%
10	Lee *et al.* (2018)	9	SHE-PWM	19.05%

11	Samadaei *et al.* (2016)	13	SHE-PWM	3.46%
12	Gautam *et al.* (2018)	9	MCPWM	——————
13	Saeedian *et al.* (2017)	9	MCPWM	11.83%

MCPWM-Multicarrier PWM, NLCPWM-Nearest Level Control PWM

Table 4.8 and Table 4.10 shows the comparison of the proposed method with the existing methods. From the tables it is observed that SHE based PSO, FSO and PSFSO presented by Panda et al produces a greater amount of harmonics when compared with the proposed system. Mahato et al, Lee et al, and Samadaei et al also implemented the SHE PWM for 9-level and 27-level inverter, but the harmonic content present in their systems is also more when compared with the proposed system.

4.9 SUMMARY

A 9-level and 27-level MMI for the elimination of certain harmonic orders(H_3, H_5, H_7 for 9-level) is developed for power quality improvement. The selective harmonic elimination method is proposed with optimization methods (GA, PSO, and BCO). The experimental investigation is carried out in the laboratory with an FPGA processor in arriving at the reduction of three possible harmonic orders for a 9-level inverter and reduction of twelve possible harmonic orders for a 27-level inverter. From the results it is observed that ,SHE-PSO method produces the output with less content of harmonics ,which is acceptable based on the IEEE519 standards. PSO has superior performance than other methods since it has an inbuilt guidance strategy which lets the solutions to obtain a useful information from the better solutions and thereby helping them to improve their own solutions. The results of PSO are faster convergence for the solutions. PSO uses memory to store the previous best solutions obtained by every candidate. This helps the candidates to recover their solutions when they get diverted to an unwanted direction. So, when the candidates start exploring an unwanted path and the quality of solutions starts degrading, they can revert and get directed to the previous stage through the P_{best} (personal best) component in the velocity term. This makes the algorithm very robust and susceptible to degradation

CHAPTER 5

DESIGN AND IMPLEMENTATION OF MODULAR MULTILEVEL INVERTER FOR GRID-CONNECTED PV SYSTEM

INTRODUCTION

The systematic use of fossil-fuel resources required an urgent quest for alternative energy sources to fulfil the daily demand. Photovoltaic power plants are gaining interests among the renewable energy sources, with high demand in the energy sector and to minimize pollution caused by excessive use of non-renewable energy (Kumar & Majid (2020)). Neither a stand-alone solar photovoltaic system nor a wind power system can however provide continuous energy due to migratory and periodic weather changes Vijay *et al.* (2020). Therefore, Grid Connected PV Systems (GCPVS) are introduced to meet the load requirement. For GCPVS, several system architectures are planned.

For GCPV power applications, four different system configurations are used: the central inverter system, the string inverter system, the multi-string inverter system and the module integrated inverter system Khan *et al.* (2020). The key benefits of using a GCPVS are: low environmental impact, can be implemented near to the customer thereby reducing the losses of transmission lines, maintenance cost is decreased since there are no moving components, and no carbon dioxide emission (Rajeev & Agarwal 2020). For small - scale distributed generator systems, like residential energy utilization, the inverter types above are used, rather than the central inverter system. Power inverters play a major role in a GCPV network. They are introduced for power transmission, synchronization and control. In grid synchronization, the system's steady state analysis and control technique play an important role.

The output of the inverter should be sinusoidal for appropriate synchronization of grid. Therefore, it is evident that the input provided to the inverter for the PV system should have lower THD, and faster dynamic response. To gain maximum power from the PV array, and inject sinusoidal current into the grid, an interface between the PV array and the grid must be used for GCPV systems. To satisfy these criteria, three types of PV configurations are mainly used Kuncham *et al.* (2020), Myneni & Ganjikunta (2020). Figure 5.1 presents the configurations of PV system.

The present work uses the two-stage configuration of PV system. The SHE-PSO method is used to generate pulses to MMI, thereby increasing the output power quality. The closed loop control is implemented for the proposed GCPVS using Proportional Integral (PI) based voltage and current control loops. The voltage control loop controls the DC link voltages from the PV sources and the Current loop controls the inverter current. Since the

performance of the controller depends on the gain parameters (Kp, Ki),

optimization techniques such as Particle Swarm Optimization (PSO), Fire Fly Algorithm (FF) ,Harris Hawks Optimization (HHO) are used to tune the gain parameters.

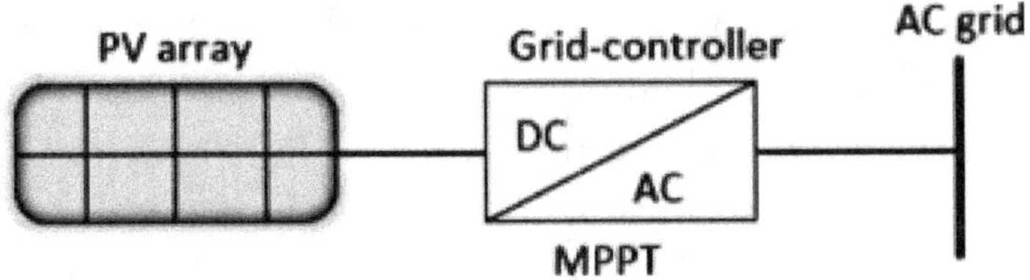

(a)

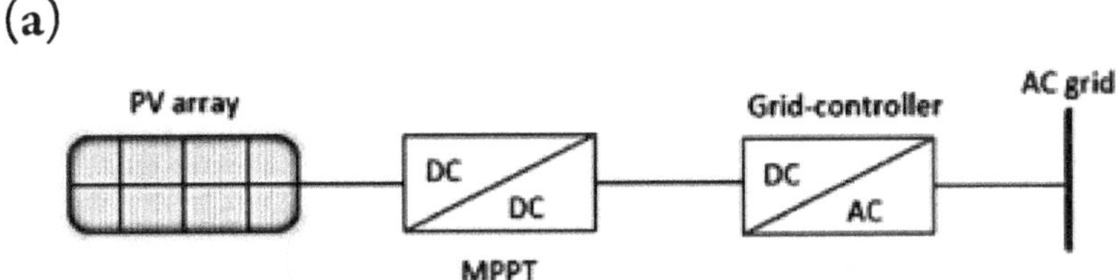

(b)

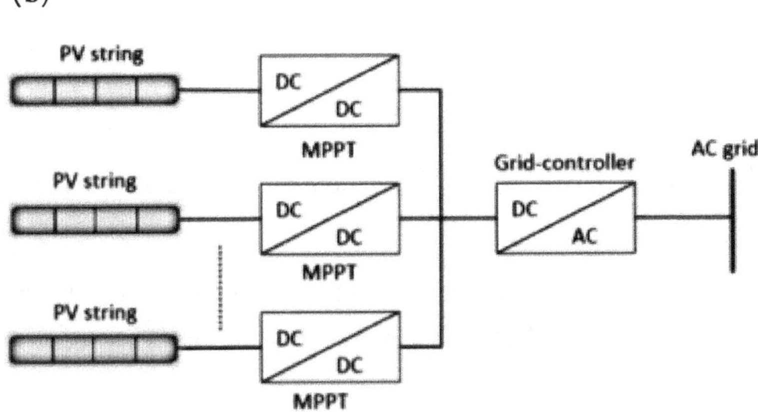

(c)

Figure 5.1 a) Single-stage, b) Two-stage, c) Multi-stage PV systems configurations

Single-phase PV systems are implemented for low-power applications of only a few kilowatts at a unit power factor (PF) with maximum power point tracking (MPPT). The Photovoltaic system gives power to the grid when the output power is greater than the load requirement. The flow of active power depends on the power produced by the photovoltaic device or the power required by the load. Figure 5.2 indicates the power flow when the unity power factor on the grid side is needed for low or medium-power applications according to the current grid standard.

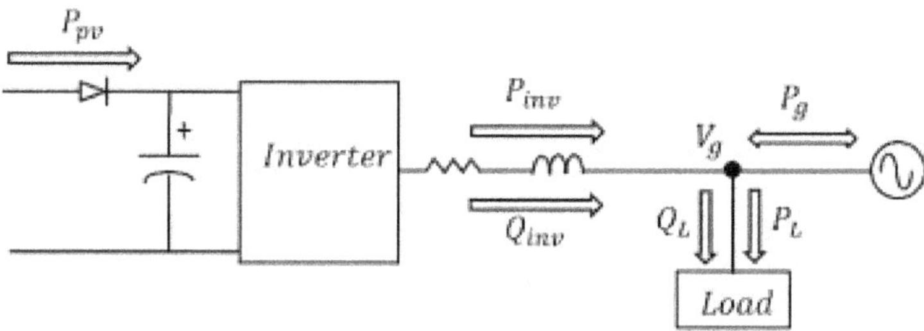

Figure 5.2 Power flow in single Phase PV system for unity power factor

In Figure 5.2 P_{pv} is the active power supplied to the inverter from the PV sources, P_{inv} and Q_{inv} is the active and reactive power of the inverter, V_g is

the grid voltage and P_g is the grid power The total active power and reactive power of a single-phase grid-connected Voltage Source Inverter (VSI), considering the dq reference frame is given by the following equations

$$P = {}_1(V\,I + V$$

–

I) (5.1)

g 2 gd d

gq q

$$Q = {}^{1}(V I - V$$

–

I) (5.2)

g 2 gd q

gq d

where, V_{gd} and V_{gq} are the d-axis and q-axis voltage components of the grid, and I_d and I_q are the d-axis and q-axis current components of the inverter. The voltage of the grid is assumed to be a zero averaged sinusoid with low harmonic components. Therefore, the q-axis component of the grid voltage $V_{gq} = 0$, then:

$$P_g = \frac{1}{2}(V_{gd} I_d) \quad (5.3)$$

$$Q_g = \frac{1}{2}(V_{gd} I_q) \quad (5.4)$$

Considering the inverter as lossless and dismissing the power loss on the filter inductor, the steady-state PV power is equal to the active power of the grid given by Equation (5.5)

$$V_{pv} * I_{pv} = \frac{1}{2}(V_{gd} I_q) \quad (5.5)$$

$$P_{pv} = \frac{1}{2}(V_{gd} I_q) \quad (5.6)$$

PROBLEM STATEMENT

Power quality standards must be established throughout the grid-connected system to meet the grid interconnection. The grid interfaced inverter are to be specifically built to validate the correct photovoltaic operation and enhance the quality of power at Point of Common Coupling (PCC).

The energy extracted from the sunlight varies periodically. The solar panel has an average voltage of 12V with a current of about 7A; it varies

between 12 - 20% to 12 + 20% throughout the day. The development of

suitable power electronic circuits is important to transform this energy into the grid as per the grid requirements Akeyo *et al.* (2020).

The goals are as follows:

- To transform the unbalanced DC voltage to balanced DC voltage

- To transform the balanced DC voltage to AC voltage

- To connect the solar PV system to the single phase grid The technological features as per the goals of the system are as follows:

- To align with the grid data, provide an inverter with 1% variation

- To supply a DC link voltage to the inverter with 1% variation

- Since the input voltage is 12V with $\pm$20% variation, it requires a high voltage transfer gain DC-DC converter

One significant problem in integrating MMI with grid is the complexity of their control for synchronization. The traditional controllers are taking a longer duration to synchronize the MMI with the grid. It is a challenging issue to control the varying output voltage of the MMI owing to changes in the solar radiation and synchronize the MMI with the grid. Moreover, the use of traditional control techniques in GCPVS leads to poor efficiency. Therefore, there is a need for implementing an advance control technique to solve this problem. The efficiency and THD of the MMI is also considered as an important issue in the GCPVS.

The PI controller is extensively used in the closed loop control system since it is simple to implement, robust in nature and it has the ability to operate under dynamic conditions Naidu *et al.* (2020). The use of PI controller in modern control techniques is limited since the performance of the controller

mainly depends on the gain parameters(Kp, Ki).

These gains are tuned to improve the performance of the controller. The Ziegler–Nichols (Z–N) method is the traditional method used for tuning the gain parameters Khan *et al.*(2020). The use of ZN method consumes a larger time and creates uncertainty in optimal selection of the parameter. This leads to poor transient performance of the controller creating instability, peak overshoots and undershoots. Hence the gains of the PI controller are properly tuned to obtain a good dynamic and steady-state response of the control system. To overcome the limitations of the conventional tuning strategy, optimization techniques are used to tune the gain parameters. The use of optimization techniques for tuning the gain parameters in GCPVS ensures a better transient response during load change, improved transient stability, improved power quality, and low overshoot Jumani *et al.* (2019).

DESIGN AND IMPLEMENTATION OF MMI BASED PV FED GRID CONNECTED SYSTEM.

The grid-connected photovoltaic system is illustrated in Figure 5.3.

It consists of a PV array, DC-DC boost converter, MMI, and the grid.

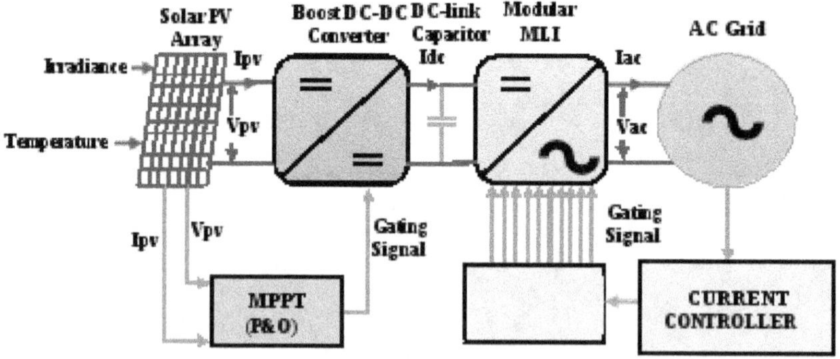

Figure 5.3 General Structure of MMI Based Single-Phase Grid Connected Solar-PV System

A PV cell produces a voltage between 0.5 and 0.8 volts based on the semiconductor and incorporated technologies. This voltage is very low to be used. Thus, several PV cells (including 36 to 72 cells) are attached in sequence to create a PV module in order to obtain benefit from the PV system. The PV modules can be interlinked in series and/or in parallel pattern to obtain a PV panel. If such PV modules are linked in series, the current would remain constant whereas the voltages are summed up.

The boost converter and the inverter are connected back-to-back. The boost converter determines the magnitude of DC-bus voltage according to maximum power generated by the PV cell. The major part of the grid system is the inverter and it is connected with the grid at the Point of Common Coupling (PCC). The interfacing of grid with the PV source is performed by the MMI. The converter, inverter design and control must be optimized to increase the overall system efficiency. The converter, inverter system should be able to produce the maximum possible power from the photovoltaic array throughout the day.

MODELING OF PV SYSTEM

A photovoltaic cell is the key component of the photovoltaic network, consisting of multiple series/parallel attached photovoltaic cells for each module. PV module signifies the main solar power generation unit. For these PV modules, the non-linear performance features of I-V and P-V depends on irradiance and cell temperature. Since the PV module contains nonlinear features, it is necessary to develop a PV module to make the exact design, function, and identification of causes of PV output degradation.

PV design is categorized into two major models; the first is a single diode, the easiest type, and the second is a two-diode type, which is efficient than a single diode. Though single diode is more common for PV modeling, there are several drawbacks as :

• Photovoltaic output with variations in temperature has high deficiencies. It avoids diffusion loss in the PV cell depletion area.

• Determination of its accuracy at low irradiance, in fact at open circuit (V_{oc})

The two-diode model is suggested to increase the precision of the PV method, boost curve fitting, and eliminate the discomfort of the single-diode model. Figure 5.4 displays the corresponding circuit of a solar PV cell. In

Figure 5.4, R_p and R_s are the shunt and series resistance of the solar PV cell.

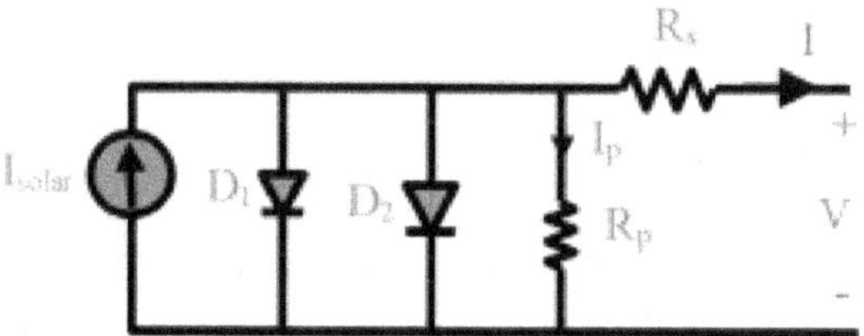

Figure 5.4 Equivalent circuit of the PV cell

Any minor variation of R_s has a significant effect on the output of PV cell. A single PV panel is made up of 24 solar cells. Each solar cell has an open-circuit voltage of 0.SV. There are 24 solar cells in a single panel. These

solar PV cells are combined in series to the produce 12V, 7A at standard test conditions (1000 W/mm2 and 2S°C). To achieve a voltage of 24V and a

current of 7A at the output from the solar PV system, two 12V, 7A modules are fixed in series.

Solar PV cell architecture involves the following equations:

$$I = I_{solar} - \mu_1 - \mu_2 - (V_s + 1^* R_s)/R_p \quad (S.7)$$

Where,

$$\mu_1 = I_{dsl} {}^* \left(e^{\frac{V_s + 1^* R_s}{\pounds_1 {}^* Vth}} - 1\right) \quad (S.8)$$

$$\mu_2 = I_{ds2} {}^* \left(e^{\frac{V_s + 1^* R_s}{\pounds_2 {}^* Vth}} - 1\right) \quad (S.9)$$

Where, μ_1 and μ_2 are the currents passing through the diodes D_1 and D_2, I_{dsl} and I_{ds2} denotes the diode saturation currents, V_s specifies the cell voltage, V_{th} specifies the thermal voltage of the solar cell, the emission coefficients $\pounds_1$ & $\pounds_2$ are related to the diodes D_1 and D_2 and I_{solar} is the current

produced by the PV cell.

$$I_{solar} = I_{cr} {}^* (I_{so}/I_{cro}) \quad (S.10)$$

Where, the solar current is indicated as I_{so}, and the irradiance current is indicated as I_{cr}, I_{cro}.

Basically, when one of solar cell in the string is having lesser amount of irradiance, it would produce smaller current compared to the unshaded cells. Since the solar cells are in series connection, every cells should have the same amount of current. Thus, the unshaded cells will impose the shaded cells to yield more current than their short circuit current and causing the shaded cells to operate at a negative voltage. Consequently, a net voltage loss occurred in the system and the shaded cells will act as a load which is absorbing power instead of producing power to the system. As a result, the partially shaded string will become open circuit and the bypass diode will be active to allow the current flow, hence the module would have lower current and power output compare to the other modules that have uniform insolation.

As the shaded cells are dissipating power and heat, the temperature of the shaded cells would increase and lead to local heating, thermal stress as well as resulting in total failure to the solar PV. In addition, power–voltage (P–V) characteristics have multiple peaks under partially shaded condition, and finding the global maximum power point (GMPP) is difficult. To alleviate these problems, several soft computing techniques integrated with P&O MPPT method is used to track the GMPP. During the partial shaded condition, the required voltage levels will be achieved with quiet reduction in voltage peak which in turn is compensated by a battery storage system. Figure 5.5 displays the PV characteristic during the partial shaded condition. This condition can be developed in the laboratory by placing cardboard on the PV panels.

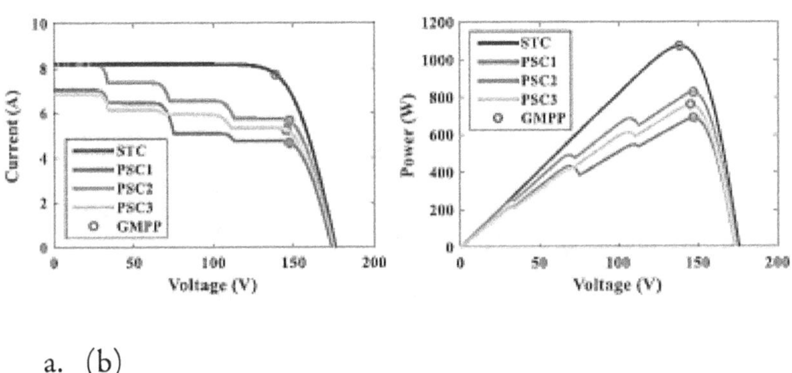

a. (b)

Figure 5.5 PV panel characteristics under four different PSCs (a) I–V characteristics (b) P-V characteristics.

DESIGN OF DC-DC BOOST CONVERTER FOR MMI

In photovoltaic systems, the DC-DC conversion has major importance because the voltage produced by the PV panels does not satisfy the load requirements. In this process, the boost converter provides maximum power to the proposed MMI.Figure 5.6 shows the struture of a DC-DC boost converter.The proposed structure uses four DC-DC boost converters. Figure 5.7 shows the struture of MMI with boost converter.

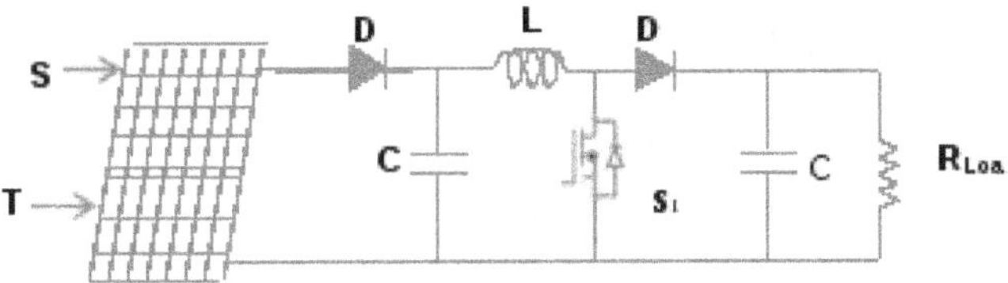

S-Solar irradiation, T-Temperature Figure 5.6 Solar PV Fed Boost Converter

The boost converter is also called a step-up converter because the name of the converter indicates a step-up in the output of DC voltage when compared with input DC voltage (Madhu Babu & Narasimharaju 2020). The boost converter includes diode,inductor, IGBT switch, and capacitor. The role of capacitor is to eliminate the distortions in the output voltage.

Inductor and Capacitor are the two important elements of a DC-DC converter which are to be modeled to get the desired output voltage. The underlying equations are used to design the inductor and capacitor.

$$V_{in} * (V_{out} - V_{in})$$

$$\underline{\qquad\qquad} =$$

$$LI * *V$$

$$(S.11)$$

L s Out

$$I_{(out)} * D$$

$$\underline{\qquad}C = \qquad\qquad\qquad {}_s * LV_c$$

$$(S.12)$$

In these equations, LI_L is the ripple value of inductor current,

$$\underline{\qquad}D = \frac{V_{out}}{V_{in}+V_{out}}$$

, LV_C

is the ripple value of capacitor current and f_s is the switching

frequency of the converter (Jagabar Sathik & Vijayakumar 2019). Equations (5.13) and (5.14) presents the formulas for calculating LI_L and LV_C.

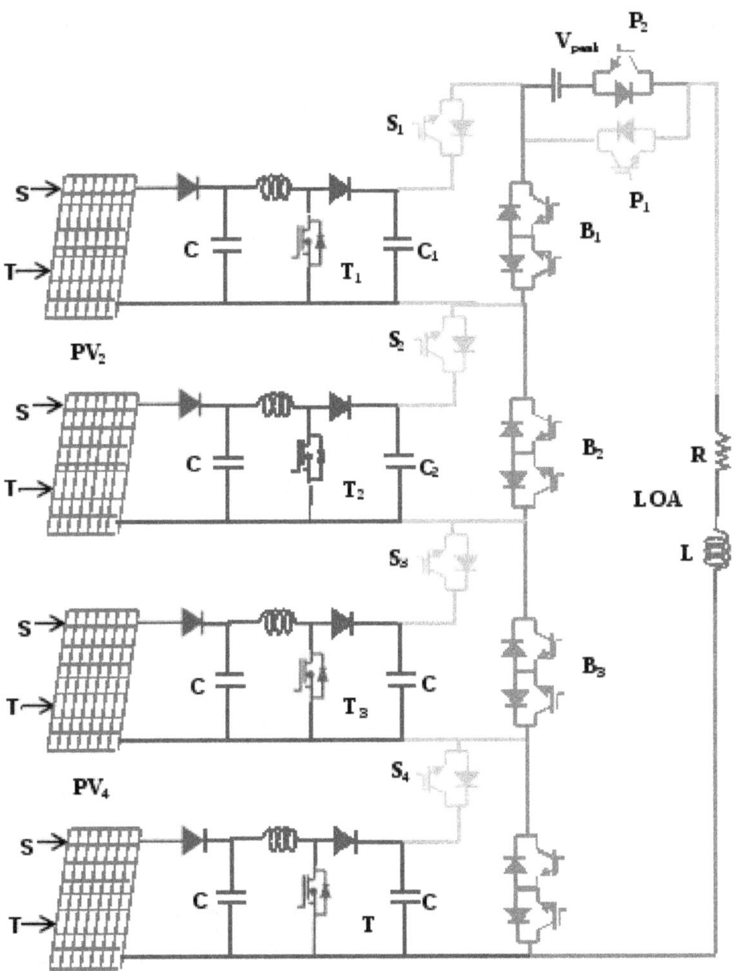

Figure 5.7 Solar PV Fed MMI

$= (2\ \%\ \text{to}\ 4\%) * I_{out(max)}$

$= (2\%\ \text{to}\ 4\%) * V_{out(max)}$

$\underline{V_{out}}$

V_{in}

$\underline{I_{out}}$

*

I_{in}

(S.13)
(S.14)

Where,V_{out} is the maximum output voltage,$V_{in}$is the input voltage from the PV module,$I_{out(max)}$is the maximum current from the PV array.

Design of Boost Converter for Symmetrical 9-Level MMI

In the proposed symmetrical MMI the input DC sources are in the ratio of 1: 1: 1: 1. Each DC source supplied by the PV source should be S0V to

achieve 200V as the peak output voltage. In general, a solar panel produces

12V under Standard Test Condition (STC). If two PV panels are connected in series the output voltage produced by the PV panels will be 24 V. Therefore to obtain S0 V from the PV source a boost converter has to be designed. The two

important elements of a DC-DC converter are the the Inductor and Capacitor which are to be modeled to get the desired output voltage.

5.5.1.1 Design of inductor and capacitor for symmetrical 9-level MMI

To obtain S0V output from the boost converter the value of the inductor is calculated using equations (5.11) and (5.13).

$$S0$$

__$LI_L = (0.03) *7* 24 = 0.43$
Substituting the value of LI_L in Equation (5.11) we get

$24 * (S0 - 24)$
_______________$= 0.43 * 1 \times 10_3 * S0 = 29mH$
The capacitor value is computed using equations (5.12) and (5.14)

7

$\_\_LV_C = 0.03 * S0 * 10 = 1V$

$\_\_\_\_\_\_\_D = V_{out}$

$V_{in} + V_{out}$

S0

$\_\_= = 0.67$

74

Substituting the value of LV_e in equation (5.12) we get

$$7 * 0.67$$

$$\underline{\hspace{3cm}}C = 3$$

$$= 4.69 \text{ mF}$$
$$1 * 10 * 1$$

Design of Boost Converter for Asymmetrical 31-Level MMI

In the proposed asymmetrical MMI the DC inputs are in 1: 2: 4: 8 ratio. Each DC source supplied by the PV source is in the ratio of 24V: 48V: 72V: 192V.To obtain the required DC voltages, four DC-DC boost converters are designed . A 12V PV input is given to the first two boost converters to obtain the 24V and 48V at the output and a 48V PV input is given

to the third and fourth boost converters to obtain 72V and 192V at the output.

To produce a 48V output from PV panel, four 12V panels are connected in

series.Table 5.1 shows the values of inductor and capacitor which are used to design the boost converter for different PV inputs.

Table 5.1 Parameters of Boost Converter

PV Input(V)	Inductor (L)	Capacitor (C)
24V	14.28mH	9.38mF
48V	10.71mH	S.6mF
96 V	S7.14mH	2.1mF
192V	42.8S mH	1.4mF

MPPT FOR PV SYSTEM USING BOOST CONVERTER

Photovoltaic performance is limited by manufacturing methods and environmental conditions. The efficiency of the PV system is typically based on the point of operation of the characteristic curves of solar modules. With solar irradiance and ambient temperature, these characteristic curves can change. To achieve the best performance, the photovoltaic systems will function at Maximum Power Point (MPP). The maximum power point changes with a change in environmental conditions. To extract maximum power the MPPT monitors the voltage and current of the PV panel.

This is achieved by changing the DC-DC converter operating period to suit the output load with the PV panels source impedance. The appropriate MPPT will usually monitor the MPP in all aspects, irrespective of the environment or the load variations. It should be easy, accurate, and implemented economically. The P&O algorithm is the most common one of all MPPT methods because of its flexibility, ease of use, and low cost. The solar photovoltaic system with P&O Algorithm is shown in Figure 5.8.

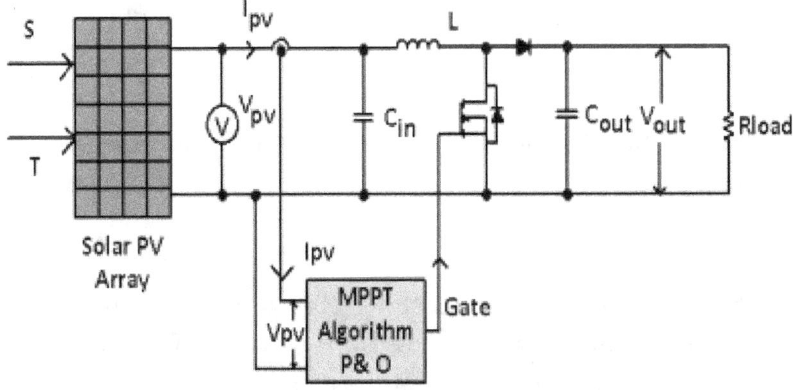

Figure 5.8 Solar-PV array with P&O MPPT controller

Perturb and Observe Algorithm

Perturbation and observation technique is the most appropriate technique for solar photovoltaic systems. The output voltage and the current from the PV system are calculated in two successive intervals. Then the power is computed using the values of voltage and current for two consecutive intervals. The rate of change of power with respect to change in voltage is

____measured as $\frac{dP}{dV}$. If the slope of $\frac{dP}{dV}$

is a positive value then the duty cycle has

____to be increased. If the slope of $\frac{dP}{dV}$

is a negative value then the duty cycle has

to be decreased. This process is continued till the maximum power point is

____reached. When the slope of $\frac{dP}{dV}$ = 0 then it is confirmed that maximum power

point is reached. Figure 5.9 shows the P-V curve of a PV module operated using the P and O algorithm. Figure 5.10 shows the flowchart of the implementation of the P and O algorithm. Table 5.2 gives the operation of P and O algorithm.

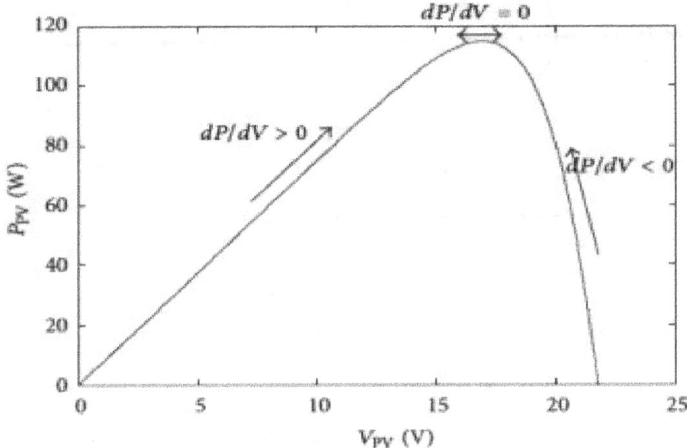

_Figure 5.9 Change in $\frac{dP}{dV}$ value of the PV module

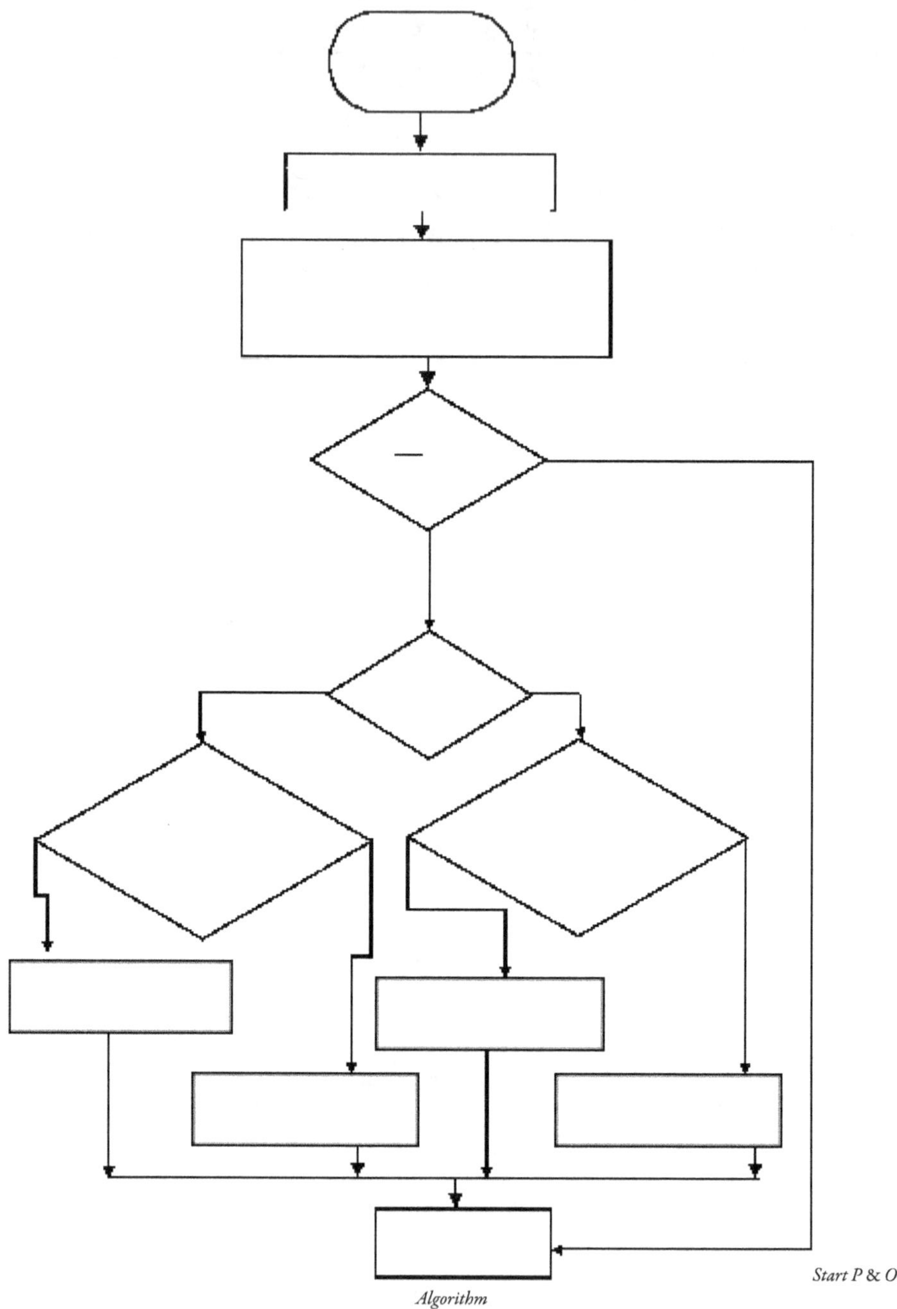

Start P & O
Algorithm
Measure: $V_{PV(k)}$, $I_{PV(k)}$

Figure 5.10 Flowchart of the P&O algorithm

Table 5.2 Functioning of P&O algorithm

dP	dV	Perturbation
> 0	>0	Increase V
> 0	<0	Decrease V
< 0	>0	Decrease V
< 0	<0	Increase V

DESIGN OF MODULAR MULTILEVEL INVERTER

The proposed MLI topology consists of 6 unidirectional switches, 4 bidirectional switches, and 5 DC sources. These components are intelligently equipped to obtain the desired output. The function of the bidirectional switch is to block the positive or negative voltages as well as allow the flow of currents in any direction. Since the common emitter configuration has fewer conduction losses and requires only one gate driver circuit it is chosen for the proposed topology.

Based on the input DC supply the proposed inverter can be operated in both symmetrical and asymmetrical modes. The proposed inverter is suitable for solar PV applications since 5 isolated DC sources are used. The fifth DC source is selected as the combination of other four individual DC sources to produce negative levels. Therefore, H-bridge is not required for obtaining the negative levels. For the proposed grid connected system a 31-level MMI is

considered and the input PV sources are taken in the ratio of 1: 2: 4: 8.

DC input of Modular Multilevel Inverter

For proper functioning of the Grid Connected System (GCS), the

_inverter input DC voltage must be greater than $\sqrt{2}V_{grid}$, where V_{grid} is the grid voltage. In this system, the grid phase voltage V_{grid} is 230 volts. Therefore, the

_minimum value of input PV voltage that has to be kept at the input of the MMI is, $V_{pv} = \sqrt{2} \times 230 = 32S.26V$. As the PV inputs of MMI are in 1: 2: 4: 8

ratio and each input source is multiplied with 24V the output voltage of the

inverter becomes equal to 360V which is enough for grid integration. Equation

A. 1S shows the calculation of input DC voltage of MMI.

$$V_{peak} = 24V + 48V + 96V + 192V = 360V \quad (S.1S)$$

The suggested inverter's maximum voltage is 360V, which is

significantly higher than the grid voltage. This ensures that the power from PV array flows continuously into the grid.

Relationship Between DC Input Power and AC Output Power of MMI

In fact, S0% of conventional losses are taken into account when

transforming the DC power to AC power. The solar PV system generates output power which is 1.5 times higher than the output power of the inverter. The DC inputs are implemented based on the output voltage of the inverter. The relation between the AC power (from inverter) and the DC power (from PV panels) is indicated in Equation 5.16.

$$P_C = {}^* P_{DC} \quad (S.16)$$

Where, denotes the conversion efficiency.

Equation (5.16) can be rewritten with the PV parameters as:

$$P_C = I_{rr} * A_m * Y_m * Y_i * p_{loss} \quad (S.17)$$

Where,

(I_{rr} is the panel irradiance in $K_{wh}/2$), A_m

is the are of the module

(m^2), Y is the module efficiency $= \left(\text{output of module}\right) * 100$, Y

is the inverter

m

efficiency $= ($ Pinv out

$), p$

$1*i$

is inverter loss $= p + p$

Pinv out+ploss

loss

edn sw

The DC power produced by a photovoltaic system should be kept constant in the suggested grid-connected network while the power delivered by the inverter to grid varies with time. Therefore, an imbalance occurs among the DC input power and the AC output power. To overcome this imbalance, capacitors are implemented in parallel to the photovoltaic network.

For the suggested MMI the DC input power is

$$P_{DC} = V_{pv1} * I_{pv1} + V_{pv2} * I_{pv2} + V_{pv3} * I_{pv3} + V_{pv4} * I_{pv4} \quad (S.18)$$

The power injected to the grid is

1 1

$$\_P_c = V_c * I_c = 2\,VI(\cos q>) - 2\,VI(\cos 2G)t + q>) \quad (S.19)$$

By taking phase shift as zero the above equation (5.19) becomes

$$P = {}_1VI - {}_1VI(\cos 2G)t) \quad (S.20)$$

$\overline{}$

C 22

_From Equation 5.20 the primary term $({}^1VI)$ specifies the normal

2

output power and the secondary term specifies the pulsating power

_(₁ VI(cos2G)t)).Since the secondary term is pulsating its frequency is double

2

the grid frequency. Considering a lossless inverter in which the PV power is

equivalent to the normal output power i.e. P_{DC}

$=_1 VI$ therefore,

_²

$$P_C = P_{DC} - P_{DC}(\cos2G)t \quad (S.21)$$

The distortions present in the pulsating power is decreased by implementing decoupling capacitance. Decoupling capacitance can be calculated using Equation (5.22).

$$C = \frac{PDC}{D\,2\,{}^{*}fi\,{}^{*}VDC^{*}}$$

$$(S.22)$$

Where, grid denotes the grid frequency which is S0Hz, V_{DC} denotes the voltage across the decoupled capacitance which will be equal to the total PV voltage from individual panels which is 360V and tJ denotes the voltage ripple across decoupled capacitance which is very low and can be neglected.

DESIGN OF LCL FILTER FOR GRID CONNECTED SYSTEM

The filter is normally linked between the inverter and the grid. The filter is used to conduct three simple functions, such as reducing high-frequency noise, shielding the transient, and translating the voltage to a current source. There are essentially three styles of filter used in the grid-connected network, these are L-type, LC-type, and LCL-type filters. For this application, the LCL filter is implemented to interconnect the inverter with the grid. Because the inverter is centered on the switching devices and the gating signals in the form of pulses must be given for the switches, the output current may contain considerable harmonic distortions which aim to reduce the power quality. The parameters for determining the values of the components are described below.

396

All the calculations are carried out using a per phase circuit (Vijayakumari *et al.* (2015)).

The inverter side inductor is sized as,

$$L_i = \frac{V_{DC\,Link}}{16 * f_s * LiI}$$

$$(S.23)$$

where, L_i is the inverter side inductor per phase, f_s is the switching frequency of the inverter $= 1kHz$, $V_{DC\,Link} = 360V$, and LiI_L is the inductor

ripple current and is calculated using Equation (5.24).

$$LiI_L = \frac{0.1 * V}{\sqrt{2}\,P}$$

$$(S.24)$$

$ph(grid)$

where, P is the power which is 11S0W, $V_{ph}(grid)$ is the phase voltage of grid $= 230V$. After implementing the above values in Equation 5.23:

$L_i = 31.82\ mH$

The inductance on the grid side is,

$$_g = 0.6 * {}_i \quad (S.2S)$$

This gives, $_g = 19.09$ mH

The capacitance value is considered as 5% of base capacitance and detrmined using,

$$C_f = 0.0S * Cb \quad (S.26)$$

_______Where, Cb = $\frac{P}{i^* V i 2}$

, P is the single-phase power = 11S0W,

w_{grid} is the rotational frequency of grid = 314.2 rad/s, V_{grid} = 230V.

The obtained capacitance for the filter is,

C_f = 3.4SμF

CONTROL OF THE SUGGESTED GRID-TIED MMI

This section discusses about the control of the GCPVS based on AMMI associated with four dissimilar PV inputs. The DC inputs of the

suggested AMMI are fed from the PV panels of magnitude V_{DC}, $2V_{DC}$, $4V_{DC}$ and

$8V_{DC}$. An effective control strategy has been implemented to inject a harmonic

less current with sinusoidal wave shape and maintain an in-phase relationship with the grid voltage. To achieve so, the separate DC inputs must be retained

identical to their reference voltage values (V^*, V^*, V^*, V^*), which are

$DC1\ DC2\ DC3\ DC4$

considered in 1: 2: 4: 8 ratio under varying irradiance condition. For this

purpose, four independent voltage regulators and one additional total voltage regulator is implemented in the proposed control system of the MMI. The architecture of the suggested grid-connected controller (GCC) is intended to tackle the subsequent characteristics:

• To control the DC inputs of the suggested AMMI independently and inject maximum power to the grid.

• In order to establish a unity power factor, the current evolved from the output of inverter is supplied to the grid with less distortions and is maintained in line with the grid voltage.

In the suggested structure, as asymmetrical PV inputs are linked to the inverter through DC links, independent DC-DC boost converters with their respective MPPT controllers are implemented independently to each PV source. These individual MPPT controllers are implemented to maintain the

DC link voltages in 1: 2: 4: 8 ratio, under different operating conditions.

The P&O MPPT algorithm is implemented to increase the output PV power for producing the reference signals (V^*, V^*, V^*, V^*) to the DC

DC1 DC2 DC3 DC4

voltage controllers.

Total DC link voltage Controller

The closed-loop voltage controller shown in Figure. 5.11 is used to maintain a minimum DC link voltage equivalent to the constant reference voltage corresponding to the irradiation level. To regulate the overall DC

voltage, the DC voltages which are taken as reference $(V^* + V^* + V^* +$

DC1

DC2

DC3

DC $V^*)$ is equated with the overall calculated DC voltages $(V_{DC1} + V_{DC2} +$

$V_{DC3} + V_{DC4})$ and the error which is obtained as the output is transmitted via the proportional and Integral controller(PI).The

PI type controllers contain gain parameters (Kp_{vl}, Kp_{il}),which are tuned using the optimization techniques

such as PSO,FFA and HHO to supply maximum reference grid current (I_{max})

gto the grid in the proposed GCPVS.For the GCC the Phase Lock Loop (PLL) generates a sinusoidal wave shaped reference grid current (I^{*}) as indicated in

Figure 5.11.

A PLL is often equipped for grid synchronization. Figure 5.12 illustrates the basic architecture of the PLL.The parts in a PLL are a Phase Detector (PD), a PI based Low Pass Filter (LF) and a voltage-controlled oscillator (VCO). The phase detector block is used to detect the phase difference between the input signal and the feedback signal. The input to the PD block is given in Equation 5.27.

$$\_V_G(t) = \sqrt{2}V_{grid(rms)}\sin(8_g) = \sqrt{2}V_{grid(rms)}\sin(w_{gt} + q_{>g}) \quad (S.27)$$

The output of the phase detector block is

$$-V_{PD}(t) = \sqrt{2}V_{grid(rms)}\sin(\theta_g)\cos(\theta^*) \quad (S.28)$$

Equation 5.28 can be expressed as

$$-$$

$$\frac{\sqrt{2}\sqrt{2}}{V_{PD}(t)} =$$

$$2\,V_{grid(rms)}\sin\left[(w_g - w^*_g)t + (\theta_g - \theta^*)\right] + 2\,V_{grid(rms)}\sin\left[(w_g\right.$$

$$\underline{\quad\quad} + w^*_g)t + (\theta_g + \theta^*) \quad (S.29)$$

Equation 5.29 consist of two elements (high frequency and low frequency). When the output of the phase detector is passed through the PI based LF block the high frequency element is removed by the low pass filter and the frequency element is obtained at the output of the LF block.

$$-$$

$$V_{LF}(t) =$$

$$\sqrt{2}$$

$$\underline{\quad} 2\,V_{grid(rms)}\sin\left[(w_g - w^*_g)t + (\theta_g - \theta^*)\right] \quad (S.30)$$

The required frequency can be obtained from the PI controller. During the stable condition the output signal from the LF block shown in

Equation 5.30 tends to zero as $w^*_g = w_g$ and $\theta^* = \theta_g$. Thus the output phase

signal is locked by giving the grid voltage as input.

Control of Grid Current and Grid Voltage Tracking

ꟼFor enhancing the inverter's output current, the reference current of the grid (I^*) which is obtained from the product of sine component of PLL and output signal of total voltage controller block is equated to the instantaneous value of the grid current (I_g)

.

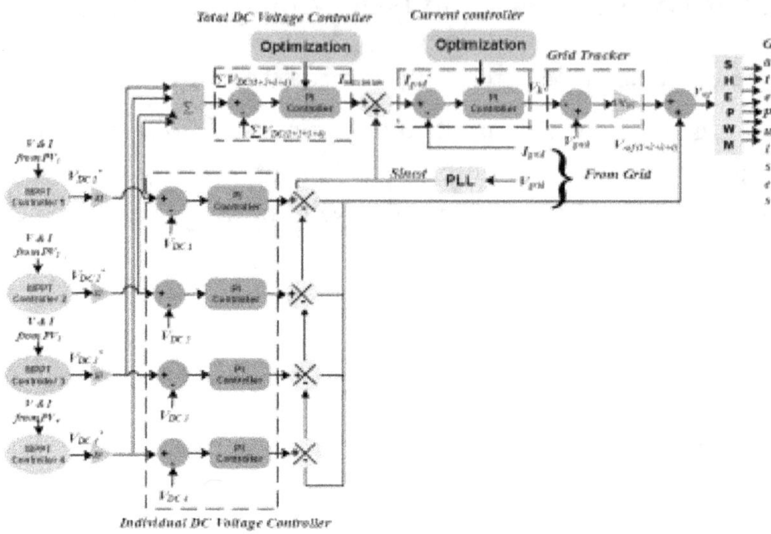

Figure 5.11 Closed-loop Control system of the proposed GCPVS

The current error which is produced in the comparator is transmitted via the PI current type controller. The current controller gains (Kp_{el}, Kp_{el}) are calibrated using the optimization techniques in such a way that the grid current is maximized and becomes equivalent to the reference grid current (I^*).

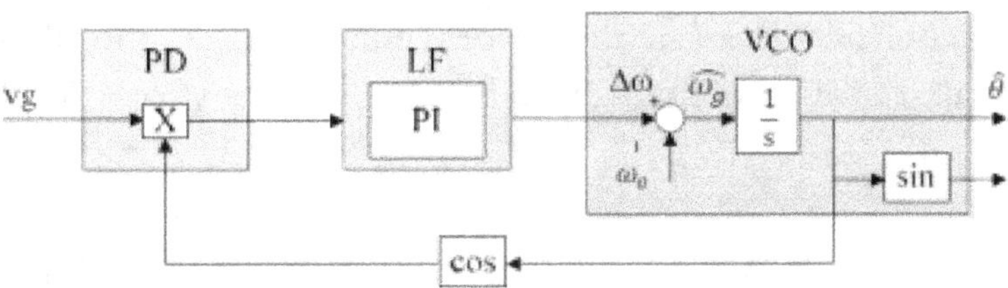

Figure 5.12 General structure of a single-phase PLL

During normal operating condition, the current controller produces the reference signal (V^*) which can be omitted from the grid voltage (V_g) for the production of reference inverter voltage (V_{ref}). This is known as the grid

tracker since the inverter's reference voltage differs based on the variations in

the grid voltage. If the voltage in the grid decreases, the difference between (V_g

$- V^*$) decreases and thus the inverter reference (V_{ref}) decreases. Therefore, the basic voltage of the inverter still continues to follow the grid voltage.

Control of Independent DC Link Voltages

For a 31-level AMMI with 4 dissimilar DC inputs (V_{DC1}, V_{DC2}, V_{DC3}, V_{DC4}), the ratios of DC sources are to be kept in 1: 2: 4: 8 ratio under different environmental conditions. Apart from the overall DC

voltage controller there are three independent voltage controllers which are essential to maintain the DC inputs V_{DC2}, V_{DC3}, V_{DC4} equal to their respective

reference voltages (V^*, V^*, V^*) as shown in Figure 5.11. Thus, utilizing

DC2 DC3 DC4

the above four voltage regulators, the individual DC voltages

(V_{DC1}, V_{DC2}, V_{DC3}, V_{DC4}) are kept at the required reference values (1: 2: 4: 8).

Hence for producing the PWM pulses to the suggested 31-level AMMI the reference inverter signal is needed. This reference signal is produced by multiplying the output of the voltage regulators with the sine component of

the grid frequency and summed up with the grid voltage (V_g).

PI CONTROLLER

A PI controller is a closed-loop or feedback system since the calculated value of the controlled parameter is fed back to the comparator. The controlled parameter is equivalent to the target value or set-point in the comparator. If there is a variation between the calculated parameter and the set- point, an error is produced. This error approaches the controller, which in return changes the final control element to return the controlled parameter to the setpoint. Negative feedback means that the difference between the calculated parameter and the setpoint is used to change the control factor such that the potential to the error is reduced. If the signal collected from the comparator is obtained by combining the two signals, there will be a positive feedback mechanism that is unreliable. The PI controller has been commonly used in all forms of feedback systems as seen in Figure 5.13.

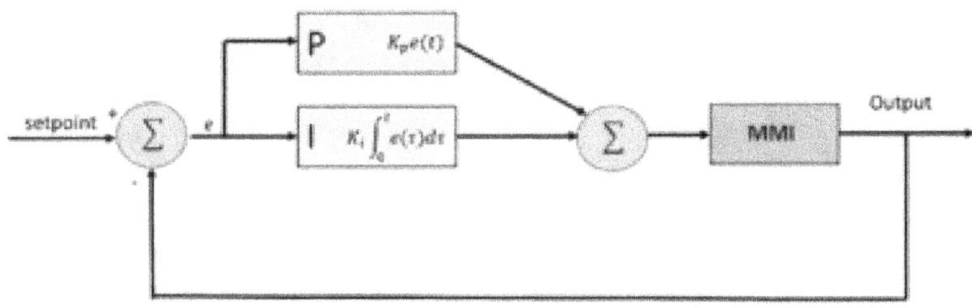

Figure 5.13 PI controller

The weighted sum of the error (the difference between output and target setpoint) and the integral value can be controlled by a PI controller. The gain of the PI controller are tuned using various methods. But if the gains of the PI controller are determined by trial and error, the steady-state error problems occur.

TUNING OF CONTROLLER

Parametric tuning and PI controller optimization can be seen as one of the most critical engineering tasks during control system authorizing to obtain the preferences for control responses. Parametric PI Controller tuning is a compromise between reaction speed and small-signal disturbance stability and tolerance to major signal disturbances. Initially, Ziegler-Nichols (Z-N) proposed new guidelines for controller tuning. This method completely uses the trial and error method which is not considered for the open loop control process. In recent studies, the author uses power and current control loops to normalize the power flow of the GCPVS. Due to its durability, easy operation, and reliability, the proportional-integral (PI) controller is commonly used for control loop control operation in GCPVS. The limitation of using a PI controller in GCPVS is their functioning, which relies solely on suitable gain

tuning to get the appropriate K_p and K_i values.

In general, these two coefficients are chosen with the "trial and error" method or with the Ziegler – Nichols (Z – N) method. The traditional method of tuning controllers is a model based on trial and error, which makes it complex to apply the PI controller in most of the issues. This approach involves uninterrupted calibration of parameter and the observation of a long time-consuming and effort-intensive response. Therefore, there is no guarantee that the result obtained would be ideal after so many variations. The Z-N method can cause high peak overshoot, large oscillations and a longer settlement time with the closed loop response of a higher order system. The two key drawbacks to conventional PI tuning approaches include high time usage and the difficulty to choose the optimal range of parameters. It can lead to a weak dynamic response of the power system and uncertainty due to significant overshoots in current and power. Therefore, appropriate tuning of PI parameters for any control system is an important task to achieve improved transient and steady- state response.

TUINING USING OPTIMIZATION ALGORITHMS

New methods in soft computing is implemented to improve the nonlinear behavior of the GCPVS to address the reported limitations of the PI controller. The application of optimization algorithms for optimal parameter selection in the grid integrated Photovoltaic system ensures low overshoots, improved power quality for consumer, smooth connection and disconnection of renewable energy system to the power system, smoother power-sharing between the grid, eased settling time, improved dynamic reaction during load changes, and enhanced energy efficiency. In addition, these intelligent search methods provide an improved approach compared to conventional analytical methods and the optimal approach to the given problem.

The basic aim of the present work is to accomplish optimal power management by avoiding traditional PI tuning techniques which are time consuming and inefficient. The results of these studies show that, compared to traditional PI tuning techniques, the PI coefficients chosen by the optimization procedures headed to an improved transient behavior of GCPVS. Many optimization algorithms such as genetic algorithms, heuristic approaches, evolutionary programming, etc. are suggested or established to date. Analysis and comparisons were mostly performed with the recent methodologies like Particle Swarm Optimization (PSO) algorithm, Firefly algorithm (FFA), and Harris Hawks optimization (HHO).

Particle Swarm Optimization

In the PSO algorithm a number of entities known as particles are used. Each particle is represented by its particle number. Each particle is associated with a vector known as the solution vector. The solution vector has a number of elements. The number of elements in the solution vector is equal to the number of variables involved in the search operation. The dimension of the search space is equal to the number of variables to be estimated.

To start with the PSO algorithm particles used in the process are initialized with random values. However, the initialization is done with values that could lie within the solution space. After the particles are initialized, each particle with its initialized values is initiated in the objective function. That particle that gives the result closest to the expected object is the best particle and its present solution is the globally best solution vector. At the end of the first iteration the suction vector of the individual particles are maintained as their personal best. Based on the personal best of each particle and the global best the individual solution vectors of each of the particles are modified.

Every particle assumes a position in the search space. At the end of each iteration the solution vectors of all the particles are shifted to new position. To shift the particles from the present position to the new position a velocity vector is added to the solution vector of each of the particle.

The velocity vector used to modify the position of the individual particles are calculated for each of the particles by considering the Individual best of each particle and the global best solution vectors. Besides two random numbers and two constants are also used in the estimation of the velocity vector. Equation 5.31 is used for the estimation of the velocity vector.

$$V^{m+1} = W * V^m + C_1R_1 * (P_{est} - {}^m) + C_2R_2 * ({}_{est} - {}^m) \quad (S.31)$$

Where V^m is the velocity of particle j at m iteration.W is the weighting function, C is the weighting factor, R is the random number between [01], P_{est} is the P_best of particle j, $_{est}$ is the global best of particle j , ^{m} is the current position of particle j at m interation.

The first element of the expression shown in Equation 5.31 is $W^* V^m$. This element is an inertia component which is necessary for driving the particle in the path it was moving before. 'w' has an important effect on the

velocity, if its value becomes smaller, then it fastens the convergence or else exploration is encouraged. The second element in Equation 5.31 is $C_1R_1^* (P_{est} - {}^m)$. It is the perceptual element that acts as the memory of the

particle. The third element is $C_2R_2^* ({}_{est} - {}^m)$. This element is responsible

for making the particle to move to the best position that the swarm has identified.

Once the speed of each particle is calculated, then the location could be updated by means of the equation of position change. Position modification equation is presented in Equation 5.32.

$$ {}^{m+1} = {}^m + V^{m+1} \text{ (S.32)} $$

Where ^{m}, $^{m+1}$ are the updated and the present search points respectively,V^{m+1} is the updated velocity.

The procedure is continued before the stop condition is reached. Figure 5.14 shows the flowchart of the PSO tuned PI controller.

In any PI controller when a command is supplied the controlled parameter moves from the initial equilibrium state to the final and expected equilibrium state. During the transient condition, the parameter under control follows a trajectory from the initial state to the final steady state. During this transition, at any instant there exists an error between the set value and the actual value. The instantaneous error changes from time to time until it becomes very close to zero and will be within the allowed tolerance. Because of the overshoot and the oscillations that exponentially decay to zero there could be positive and negative errors. In a discrete system the sum of the instantaneous error adds up to the integration of these errors. The Equations from 5.33-5.36 present the integral time absolute error (ITAE), integral time square error (ITSE), integral absolute error (IAE), and integral square error (ISE).

$$I E = \int e(t)^2 * dt \quad (5.33)$$

$$IAE = \int |e(t)| * dt \quad (5.34)$$

$$ITAE = \int t * |e(t)| * dt \quad (5.35)$$

$$IT E = \int t * [e(t)^2] * dt \quad (5.36)$$

These Equations are known as the performance indices which are most commonly used as the fitness functions. The performance of the PI controller depends on the performance indices. In the proposed work the ITAE is advantageous than the other performance indices since it has a reduced settling time and peak overshoot. Therefore, it is used as the fitness function for all the optimization techniques.

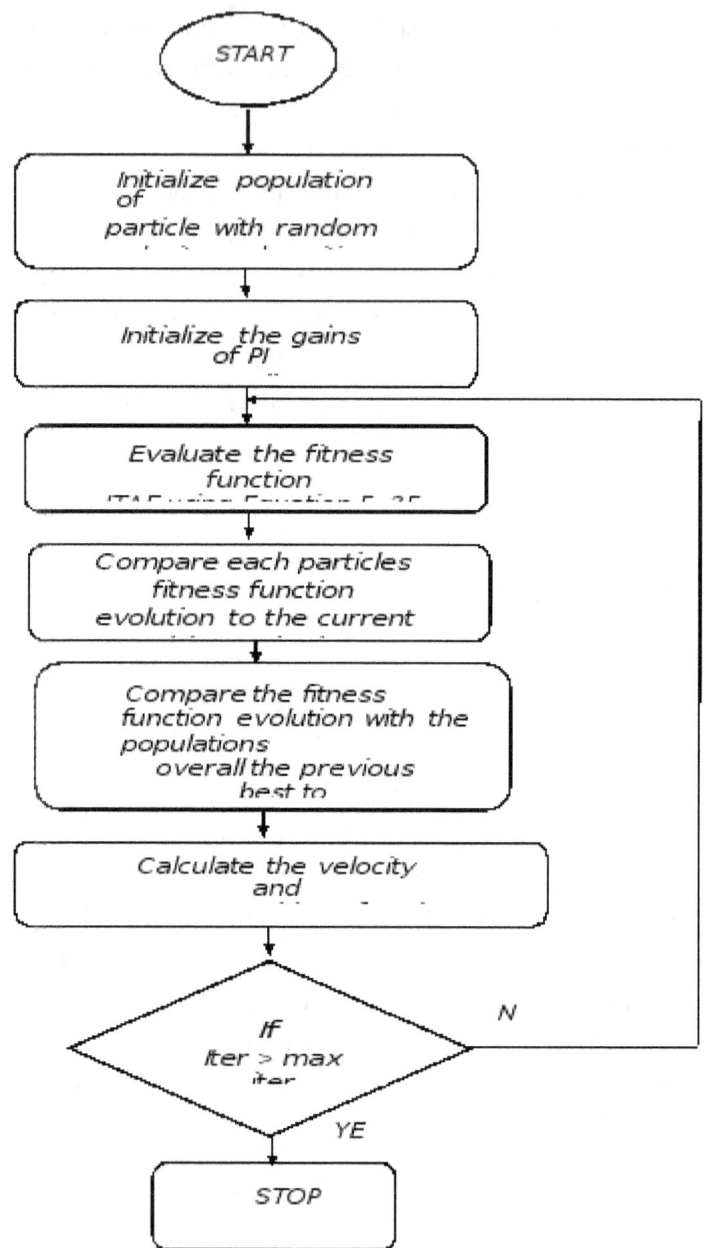

Figure 5.14 Flowchart of PSO Tuned PI Controller

Firefly Algorithm

The firefly algorithm (FFA) is focused on the fireflies' normal behavior. Those involve the usage of a community of insects to lure, hide, or

contact predators using their bioluminescence. The luminosity (I) decreases

with increasing distance (r). The light intensity is measured in order to optimize

the algorithm in a manner equal to its cost function. Three essential rules are used for a better explanation of the FFA.

- Attraction among the normal and brighter fireflies take part without considering their gender

- Attraction is approximately equivalent to the luminosity and reduces as their distance gets bigger.

- Attraction is proportionate to luminosity and increase as its distance decreases

- The luminosity is measured by the fitness value

There are two essential details to be defined for the implementation of the Firefly algorithm. These are the difference of light intensity (I) and the

composition of attraction (�^{f}). The luminosity of the objective function

decides the attractiveness of fireflies. In a specific instance, luminosity (I) at a given place depends on its position x which is illustrated in Equation 5.37.

$$I(x) \, a \, (x) \quad (S.37)$$

Light intensity and attractiveness

The β_f is an attractive factor that should be analyzed or seen by various fireflies. In general, β_f varies from 0 to 1 ,where it decides in what way the fireflies are attracted to each other. If the β_f value is decreased, the desire to step towards a luminous firefly will be diminished. The value of β^f and a random factor influences the firefly to step towards a luminous firefly. Therefore, the β^f value relies on the distance (r_i) between the ith firefly moving

towards the jth luminous firefly. In the case of a reduction in light intensity with

greater distance from the source, the attractiveness value (β^f) differs based upon the level of adsorption. The intensity of light shown in Equation (5.38) thus differs based on the inverse square distance r.

$$I(r) = \frac{I_s}{r^2}$$

(S.38)

The intensity of light changes with r in a medium provided with fixed
light absorption coefficient γ shown in equation (5.39).

$$I = I_o\, e^{-\gamma r^2} \quad (S.39)$$

Where, I_o is the actual intensity of light.

The Gaussian model is formed to prevent singularity at $r = 0$ in

Equation (5.39), by integrating the impact of the inverse square law with absorption as presented in Equation (5.40).

$$I(r) = 1 + \gamma r^2$$

$$(S.40)$$

The neighbouring fireflies will later calculate the attractiveness (β) which is proportional to the light intensity.

$$\beta^f = \beta_0 \, e^{-\gamma r^2} \quad (S.41)$$

where, β_0 is the attractiveness at $r = 0$.

This function is estimated by:

$$�^{f}(r) = \frac{�_{o}}{1 + yr_{2}}$$

(S.42)

The travelling of a firefly to meet a luminous firefly is solved by

$�^{f}(r)$ and its random element. The random element is a main element for all

heuristic methods, which makes the algorithm to get away from local optimums.

Distance

The distance x_i,x of any two i, j fireflies, is the Cartesian distance shown in Equation (5.43).

$$r_i = Ix_i - x\, I = \underset{k=l}{\overset{d\ 2}{L}} \left(x_{i,k} - x_{,k}\right)$$

(S.43)

Location update

The location of fireflies is modified when the attractiveness of firefly i to each another is greater than the luminous firefly j.

Where, $(x_{i,k})$ is the kth element of the three-dimensional coordinate x_i of ith firefly. The motion of

a firefly i towards a more luminous firefly j in a specific time (t) is decided by

Equation (5.44).

$$X_{t=1} = X_t + \blacklozenge \; e_{-yr2i \; t \; t} \; 1$$

$$-$$

(S.44)

i i o

$$(X_i - X) + a(rand -)$$

$$2$$

Where, a is the variable regulating the quantity of randomness. The

random variable, a ranges from 0 to 1. Figure 5.15 shows the flow chart of Firefly tuned PI controller.

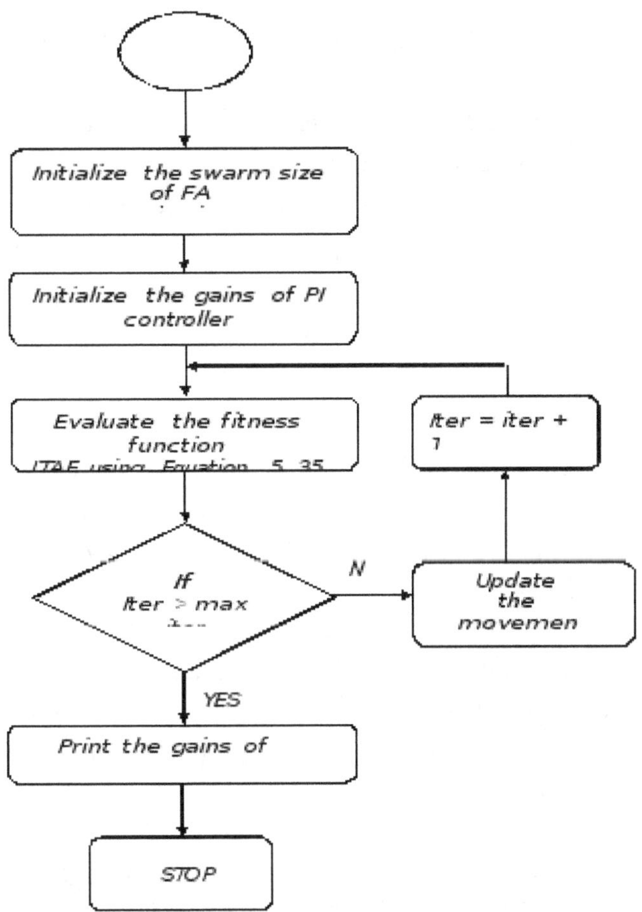

Figure 5.15 Flowchart of FFA tuned PI controller

Harris Hawks Optimization (HHO)

HHO is a modern, natural meta-heuristic optimizer inspired to find the food source by Harris hawks. Several hawks in an organized manner strike prey to astonish it (exploration phase). Regarding the possibility of escape and flight for hunting, the Hawks may make several fast dives, which surprise and make the prey tired (the exploitation phase). The HHO algorithm, which is primarily focused on the flowing energy of the prey, can switch its step from exploring to exploiting and then move between different ways of exploitation. The energy of the prey can be drastically reduced through the running phase. Figure 5.16 illustrates the stages of HHO algorithm. The energy of the prey can be described as:

t

$$\_A = 2A\,[1 - T]\ (S.4S)$$

Where A indicates the active and operating energy of the prey for each iterative process in the algorithm, T is the overall number of iterations and A is the initial energy of the prey. The developer should assign the value of

A. In this method, for every iteration, A changes randomly from (-1,1). When the value of A diminishes from 0 to -1 the prey gets exhausted and if the value of A rises from 0 to 1 the prey is reinforced. The energy of the prey

to escape from Harris hawks tends to be weakened throughout the iterations. When IAI 2: 1, the hawks are looking for different locations to find the location of the prey. This is known as the exploration process. The HHO performs the exploitation process while IAI < 1. Qais *et al.* (2019) outlined

four approaches namely Soft attack, Hard Attack, Soft Attack with gradually rapid dives, and Hard Attack with progressive fast dives for accurately modelling the method of striking the target depending on its moving condition and the hunting styles of the Harris Hawks .

The hawks can do a hard or soft attack to capture it, depending on the escape actions of the prey. In this scenario, the Hawks must besiege and strike the prey from different angles and this phase relies on the retained strength of fleeing from the prey. After several repetitions and attempts, the energy of the fleeing target is slowly lost. So, it is time for the hawks to accelerate the attacking process and they attack the exhausted prey and capture it or enter the final global response.

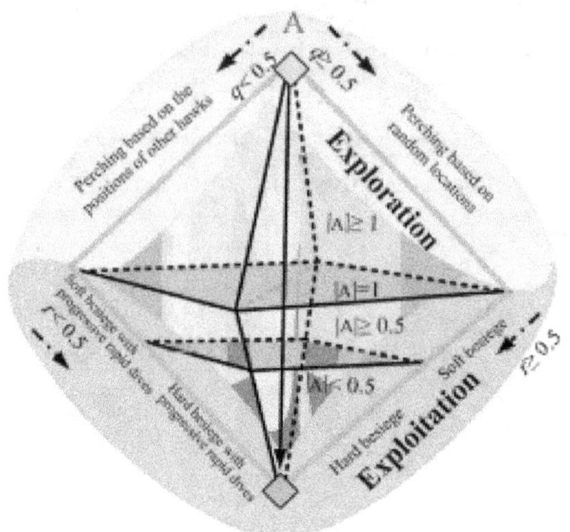

Figure 5.16 Logical Diagram Interpreting the stages of HHO algorithm

5.13.1 Step by Step Procedure for Implementation of HHO

Step 1: Start generating the inhabitants of Harris hawks between the upper and lower control parameters as formulated:

$$X(N, d) = rand(N, d) * (UB - B) + B \quad (S.46)$$

Where N is the maximum hawks representing the search agents, d is the varying attributes, UB is the upper bound parameters, and LB is the lower bound parameters.

Step 2: The solutions provided are seen as the location of the hawks, as shown in the matrix:

$$x = \begin{matrix} x_{1,1} & x_{1,2} & . & x_{1,d} \\ x_{2,1} & x_{2,2} & . & x_{2,d} \\ & & & \\ x_{n,1} & x_{n,2} & . & x_{N,d} \end{matrix} \quad (S.47)$$

Step 3: Computing the fitness (suitability) of each position of hawks as follows:

$$F_{fit} = [\ 1, 2, 3, \cdots N]_T \quad (S.48)$$

where, T is the maximum iterations

Step 4: Procuring the position of the prey based on the best fitness

Step 5: In the HHO method, hawks may be moved from exploration to exploitation, and transitions between various activities depend on the escaping energy of the prey. The prey's energy is dropped significantly during the flee. The prey's energy can be described as:

t

$$\_A = 2A\,[1 - T] \quad (S.49)$$

Where A signifies the prey's, fleeing energy and A_0 is the energy of primary state.

Step 6: For the case of lAl 2: 1, the motion of hawks is to be searched and

detected randomly. The modified position of hawks can be computed as follows:

$$x_R(t) - r_1 l x_R(t) - 2r_2 x(t) l\ q\ 2:\ 0.S$$

$$x(t + 1) = \{(X$$

$$(t) - x\,(t))\ -r\,(\ B + r\,(UB - B))\ q < 0.S \quad (S.S0)$$

RBm34

where $x(t + 1)$ signifies the location vector of hawks in the subsequent iteration t, $X_{RB}(t)$ is the location of rabbit(prey), $x(t)$ is the present location vector of hawks, r_1, r_2, r_3, r_4 and q are the random numbers ranging from (0,1), which are modified for every iteration, $x_R(t)$ signifies the arbitrarily chosen hawks from the present inhabitants and x_m denotes the average hawk's location of the present inhabitants. The average location of hawk's is presented in Equation (5.51).

$$\frac{1}{N} X_m(t) = L \; x_i(t)$$

$$i=1$$

$$(S.S1)$$

Where, $x_i(t)$ specifies the position of every hawk in every iteration
t
and N denotes the overall hawks.

Step 7: It signifies the striking on the prey.

For lAl < 1 and r = R(0) , the hawks hunt the prey with four recommended approaches which can be computed as follow:

1. This approach arises when r 2: 0.S and lAl 2: 0.S. The modified position of hawks can be computed as follows:

$$x(t + 1) = Lx(t) - AlJ \; X_{RB}(t) - x(t)l \; (S.S2)$$

$$Lx(t) = X_{RB}(t) - x(t) \; (S.S3)$$

where, $Lx(t)$ is the variation related to the location vector of the target and the present position in iteration t, J = 2(1 - rs) it signifies the arbitrary jumping energy of the target and rs is a arbitrary number in (0,1). In

every iteration, the Jth element varies randomly to simulate the behavior of prey activities.

1. This approach arises when r 2: 0.S and lAl < 0.S. In this situation, the modified position of hawks can be computed as follows:

$$x(t + 1) = X_{RB}(t) - A|Lx(t)| \tag{S.S4}$$

1. This strategy occurs when r < 0.S and lAl 2: 0.S. This approach is

more sensible than the preceding approach. The target has sufficient energy to abscond from the hawks. The motion of hawks tracks the levy flight (LF) idea which defines by an actual zigzag movement to grab the target. The subsequent hawks' location can be assessed by the Equation. (5.55):

Y =X$_{RB}$(t) - AlJ X$_{RB}$(t) - x(t)l (S.SS)

Corresponding to the patterns based on LF idea the Hawks will dive in as stated in Equation. (5.56)

Z = Y + x F (D) (S.S6)

Where D denotes the challenge dimension and S is the arbitrary vector by size 1 × D and the LF operation is determined as:

F(X) = 0.01 *

U*

———⌐

lVl$_B$

(S.S7)

=

ı

8(1 + ,9) * sin n,9

2

——————————————1+ ,9 ,9 -1

8(2)* ,9 * 2(2)

where, U and v are the arbitrary numbers ranging between (0,1). ,9 is

fixed to 1.5 which is a constant value. The approach of modifying the location of hawks can be expressed as:

$$x(t + 1) = \{Y \; i \; F(Y) < F(x(t))$$
$$Z \; i \; F(Z) < F(x(t))$$

$$(S.S8)$$

1. This strategy occurs when r < 0.S and lAl < 0.S. The target or

rabbit or prey has very less energy for absconding and is at far distance from dangerous locations. The hawks try to lessen the space among their positions and the required target. The modifying position of hawks can be computed using Equation (5.58) and (5.59).

$$Y = X_{RB}(t) - A lJ X_{RB}(t) - x(t)l \quad (S.S9)$$

Step 8: Repeat this process from step (3) to step (7) until all iterations are done.

Step 9: Finding the best solution (the target solution).

Figure 5.17 shows the flowchart of HHO tuned PI controller. The integral time absolute error is used as the fitness function by the HHO algorithm to obtain the parameters of the PI controller of the total voltage controller block and the current controller block of the proposed control system. The fitness function of HHO method for obtaining the gains of the total voltage controller

block and the current controller block is given in Equation (5.60) and (5.61)

$$ITAE = J t * le(t)l * dt = J lV_{DCref} - V_{DCaet}(t)l *t * dt \quad (S.60)$$

$$ITAE = J t * le(t)l * dt = J II_{gridref} - I_{gridaet}(t)I * t * dt \quad (S.61)$$

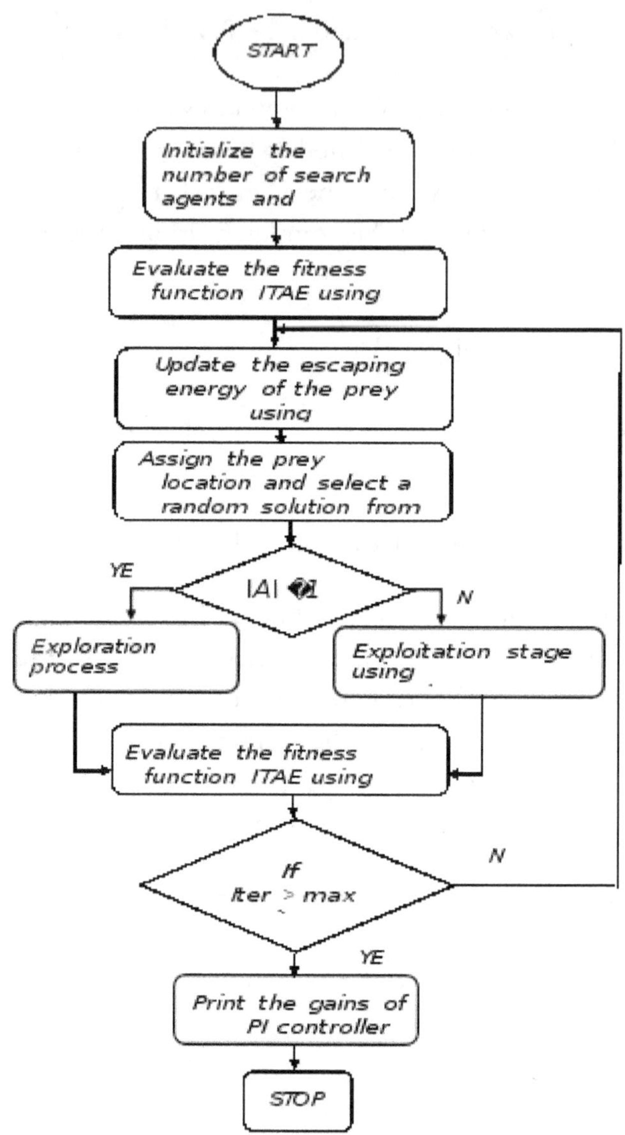

Figure 5.17 Flowchart of the HHO tuned PI controller

5.13 RESULTS AND DISCUSSION

This section discusses the simulation results of a single-phase PV fed MMI based GCPVS using PSO-PI controller, FF-PI controller, and HHO-PI controller. The usage of the proposed control technique to improve the power quality and to inject sinusoidal current to the grid is simulated in MATLAB. In the proposed control technique, the grid-connected PV system is demonstrated and the systems power quality is analyzed. For the different conditions of the grid, the adaptive control of the proposed technique delivers sinusoidal current into the grid. At the grid side, the balanced and unbalanced load variations are analyzed in the viability of the proposed system and the related execution for other solution techniques is investigated. The usage parameters such as nominal grid voltage, filter inductance, filter capacitance, grid frequency, and grid

protection are elucidated in Table 5.3. The switching frequency (f_s) is selected

as 1kHz for both hardware kit and for MATLAB simulation.

Table 5.3 Specifications of PV Fed Grid Connected System

438

PV FED MMI

Nominal grid voltage (V_{rms})	$230V_{rms}$
DC Voltage (V_{DC})	360V
Switching Frequency($_s$)	1kHz
Supply Frequency( )	S0Hz
Inverter Filter inductance ($_i$)	32 mH
Grid Filter inductance ($_g$)	20 mH
Filter capacitance (C)	3.4SµF
Grid frequency (ω)	2n * S0 rad/sec
Grid resistance (R_g)	0.0S.n

Figure 5.18 displays the input PV sources for the proposed 31-level AMMI. Figure 5.19 shows the total DC link voltage of the MMI. Figure 5.20 shows the output voltage and current of the proposed 31-level MMI. A PI controller is implemented for the closed loop control system and its dynamic

performance is improved by tuning the gain values (K_p, K_i) with optimization

techniques such as PSO, FFA, HHO.

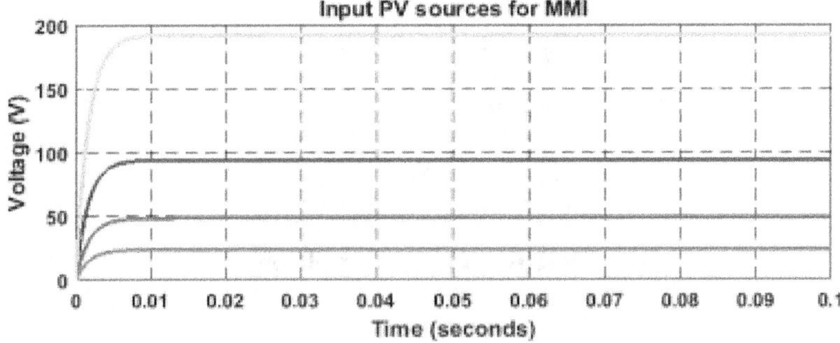

Figure 5.18 Input PV sources for 31-level AMMI

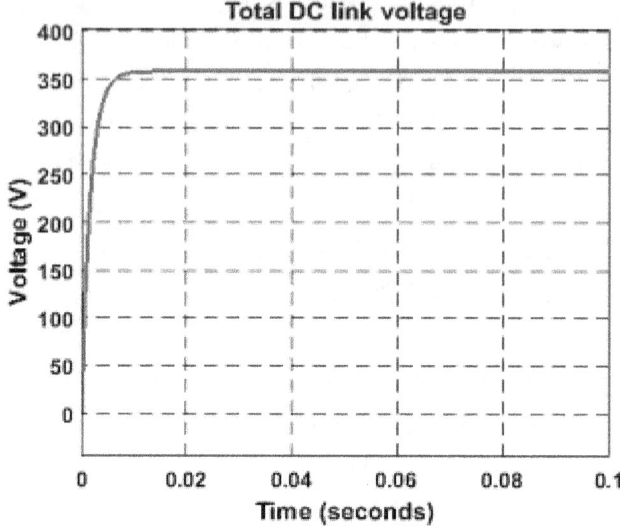

Figure 5.19 Total DC link voltage

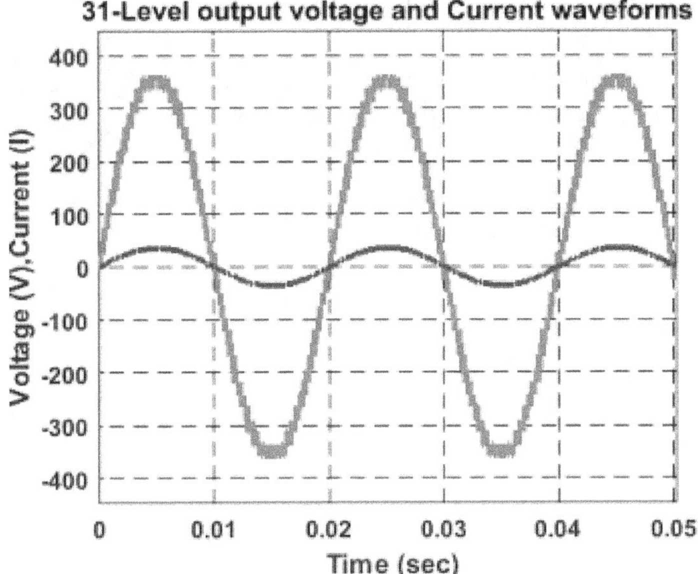

Figure 5.20 Output voltage and current of 31-level MMI

5.13.1 Implementation of PSO-PI Controller

Table 5.4 shows the parameters of PSO required to tune the PI controllers of the total voltage control block and the current control block. The integral time absolute error is chosen as the fitness function and during optimizing the fitness function, the PSO algorithm provides the gains of the two PI controllers. It is found that, the finest gain values of the PI voltage type

(K_{vp}, K_{vi}) and current type (K_{ep}, K_{ei}) of controllers are obtained after twenty

iterations. The finest gain values of the voltage and current controllers are

$K_{vp} = 0.41$, $K_{vi} = 0.20S$ and $K_{ep} = 1.1$, $K_{ei} = 0.294$ respectively. Figure

5.21 presents the grid voltage, grid current and current THD of the proposed GCPVS using the PSO-PI controller. It is observed that the current injected into the grid is in phase with the grid voltage and it has a THD of 5.97%.

Table 5.4 Parameters of PSO

Parameters	Value
warm size	1S
Maximum iterations	20
Acceleration constant C_1	0.S
Acceleration constant C_1	1.S
Initial inertia weight w_{max}	0.9
Final inertia weight w_{min}	0.4

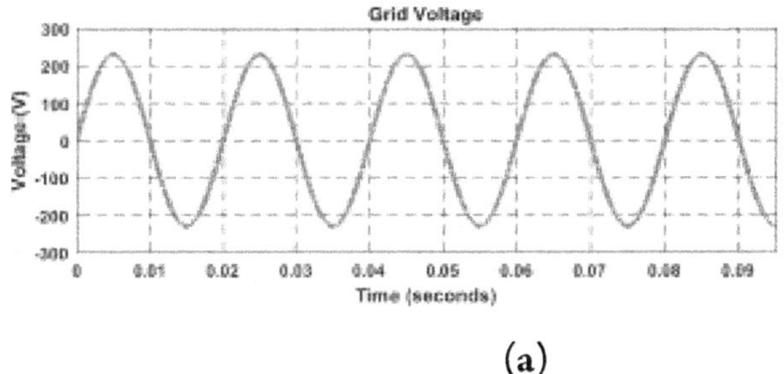

(a)

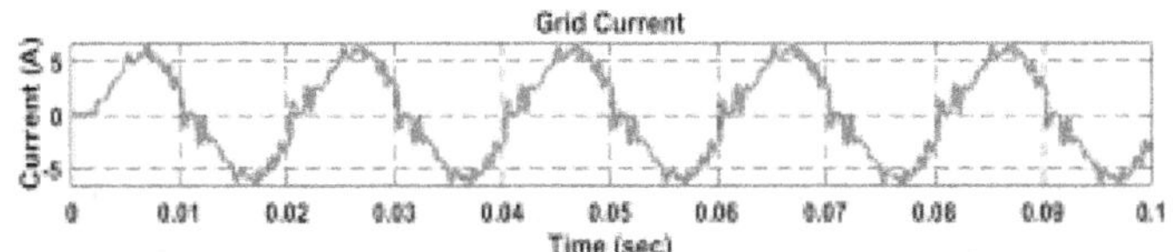

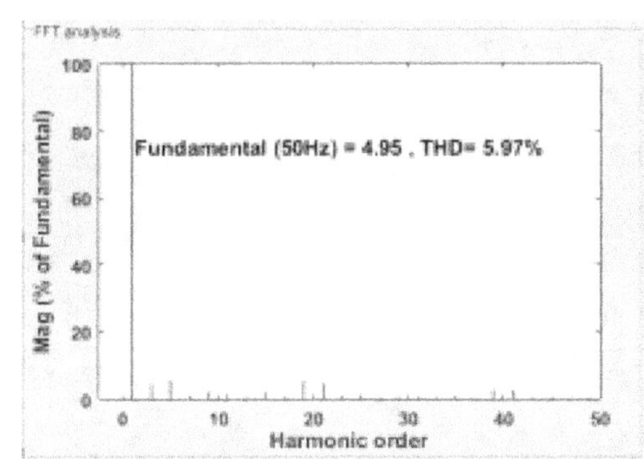

a.

b.

Figure 5.21 Simulation Result of a) Grid voltage b) Grid current c) FFT analysis of Grid current using PSO-PI controller

5.13.2 Implementation of FFA-PI Controller

In the proposed GCPVS, fire fly algorithm (FFA) is used to tune the gain values of the total voltage and current control block. Table 5.5 shows the parameters of FFA required to tune the PI controller. The cost function is chosen as the integral time absolute error since this performance parameter requires less settling time and has less overshoot.

In the process of optimizing the cost function, the FFA gives the best gain values of the PI voltage type (K_{vp}, K_{vi}) and current type (K_{ep}, K_{ei}) of

controllers .It is found that after thirty iterations the best gain values are $K_{vp} =$

0.73, $K_{vi} = 0.1S7$ and $K_{ep} = 0.7S$, $K_{ei} = 0.073$ respectively. Figure 5.22

presents the grid voltage, grid current and current THD of the proposed GCPVS using the FFA-PI controller. It is observed that the current injected into the grid is in phase with the grid voltage and it has a THD of 2.47%.

Table 5.5 Parameters of FFA

Parameters	Value
Number o ire lies (n)	20
Randomization actor (a)	0.S
Initial attractiveness(β)	0.2
Absorption Coe icient(y)	0.S
Maximum iterations	30

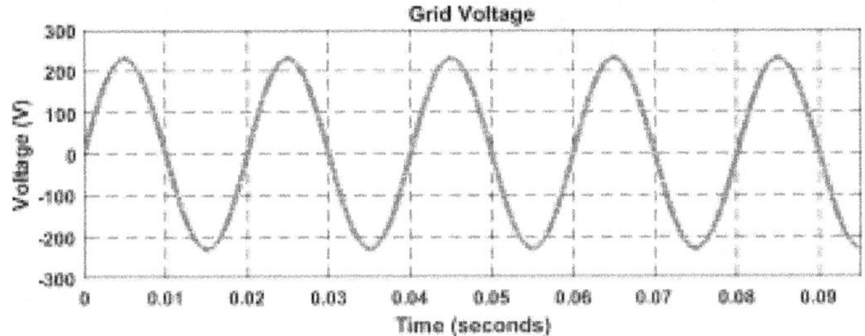

(a)

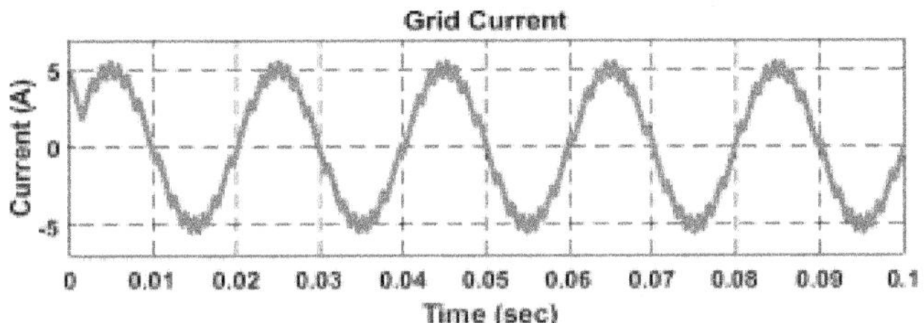

(b)

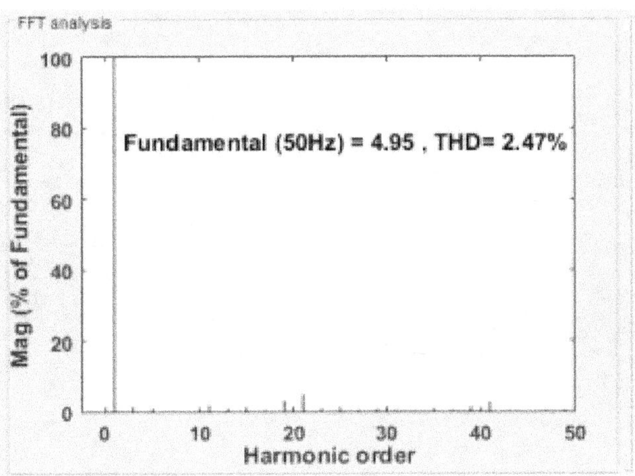

(c)

Figure 5.22 a) Grid voltage b) Grid current c) FFT analysis of Grid Current using FFA-PI controller

5.13.3 Implementation of HHO-PI Controller

In the proposed GCPVS, Harris Hawks Optimization (HHO) is used to tune the gain values of the total voltage and current control block. Table 5.6 shows the parameters of HHO required to tune the PI controller. The objective function is chosen from Equation 5.60 and 5.61 which is the integral time absolute error. Equation 5.60 is used as the objective function for the HHO-PI voltage type controller and Equation 5.61 is used as the objective function for the HHO-PI current type controller.

Table 5.6 Parameters of HHO

Parameters	Value
Number o search agents	7
Energy o the prey(A)	[0 2]
Constant (,9)	1.S
Maximum iterations	30

In the process of optimizing the cost function, the HHO gives the best gain values of the PI voltage type (K_{vp}, K_{vi}) and current type (K_{ep}, K_{ei}) of controllers .It is found that after thirty iterations the best gain values are K_{vp} = 0.93, K_{vi} = 0.091 and K_{ep} = 0.9S, K_{ei} = 0.092 respectively.

Figure 5.23 presents the waveforms of the grid voltage, grid current ,active and reactive power using HHO algorithm.

Figure 5.24 shows the current THD of the proposed GCPVS using the HHO-PI controller. From Figure 5.23 it is observed that the HHO controller injects a sinusoidal current to the grid and maintains the grid current inphase with the grid voltage.

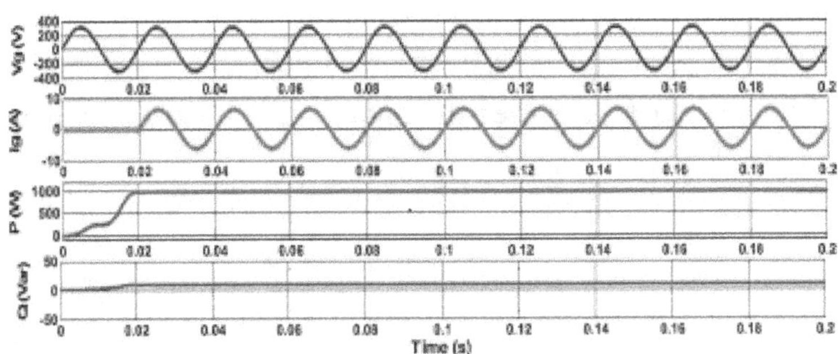

Figure 5.23 Output Waveforms of Grid Voltage, Grid Current, Active and Reactive Power Using HHO-PI controller

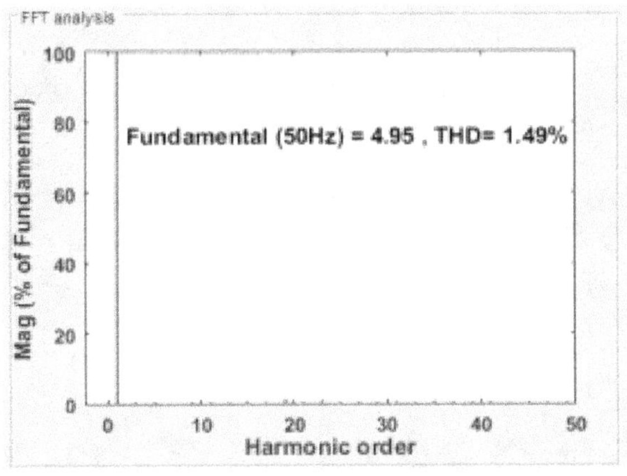

Figure 5.24 FFT analysis of the grid current using HHO

During a stable condition, the suggested inverter contributes 1 kWp of active power to the grid. The THD of the grid current during stable condition is determined as 1.49%.

5.13.3.1 Functioning of HHO in transient condition

The suggested grid-connected MMI is evaluated with different loads. The inverter injects the active power (P) to the grid when the load is inductive, leaving the reactive power (Q) zero. On the other hand, the inverter absorbs reactive power from the grid when the load is capacitive. The function of the modular multilevel inverter is to contribute constant active power to the grid when there is a change in the reactive power. In the simulation, to grasp the idea of a continuous contribution of active power to the grid the reactive power is switched from 500VAR to -500VAR at 0.1second.

The transient state waveforms are presented in Figure 5.25. From Figure 5.25 it is indicated that the current in the grid remains constant even though the power factor changes from 45o lagging to leading. Due to the fast acting of HHO-PI controller at 0.1second during the transient condition the grid current remains sinusoidal.

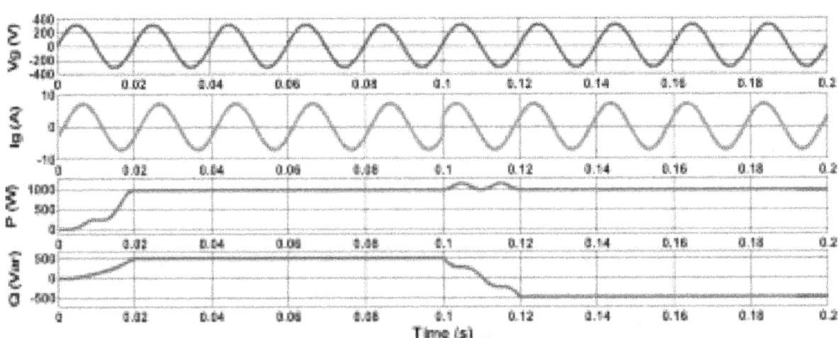

Figure 5.25 Output Waveforms of Grid Voltage, Grid Current, Active and Reactive Power during transient condition Using HHO- PI controller

The performance of the MMI and the proposed controller is examined in voltage sag and swell situations. Figure 5.26 displays the voltage sag waveform. A voltage sag is produced at 0.1second with a reference current of 5A. Figure 5.26 indicates that the current in the grid is not interrupted and is lying in phase with the voltage of the grid during the voltage sag condition. Moreover, the sag voltage is observed as 210V.

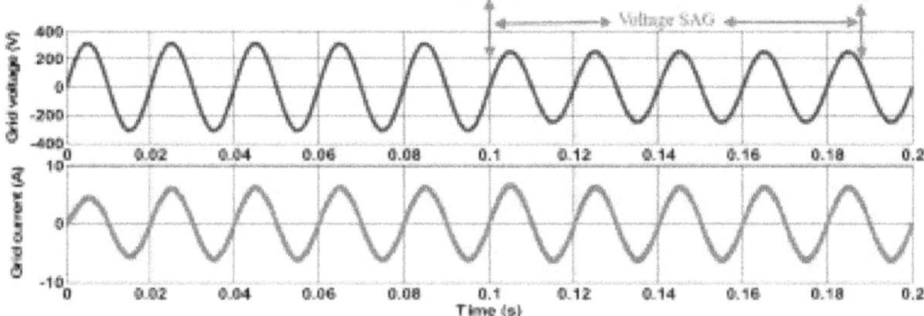

Figure 5.26 Performance of HHO-PI controller during Voltage Sag

The voltage swell is produced by triggering large capacitor stack, sudden load rejection, single ground fault line, and load de-energization. Figure

5.27 displays the system's output in voltage swell situation. When the load is disconnected abruptly at 0.1second in the proposed configuration, the supply voltage rises from 230V to 375V. Figure 5.27 indicates that the grid current is constant over a period of time and is in phase with the voltage of the grid. The voltage swell is balanced throughout the process by the help of grid control system. The grid control system keeps the current of the grid in sinusoidal wave shape . If the swelling of the voltage continues unchecked, overheating occurs and it gradually leads to destroying the inverter components.

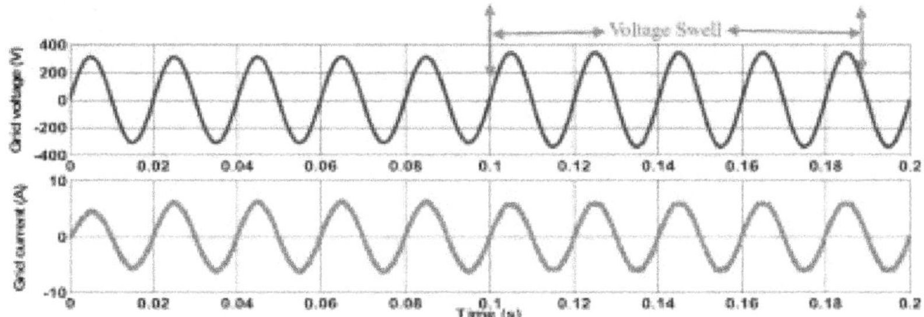

Figure 5.27 Performance of HHO-PI controller during Voltage Swell

Table 5.7 shows the THD and gain parameter values for the PSO-PI controller, FFA-PI controller, HHO-PI controller.

Table 5.7 Gain Parameters and THD Value of the proposed Grid Connected System

Controller	Gain Parameters				%THD	
	Voltage Controller		Current Controller			
	K_{vp}	K_{vi}	K_{cp}	K_{ci}	Simulation	Exper
PSO-PI	0.41	0.205	1.1	0.294	5.97	-
FFA-PI	0.73	0.157	0.75	0.073	2.47	-
HHO-PI	0.93	0.091	0.95	0.092	1.49	2.03

In contrast with the other techniques, the proposed HHO-PI controller gives less oscillations, smaller gain values, a better dynamic response with minimum overshoot, a better THD and less transient period as presented in Table 5.7.

5.13.4 Experimental Verifications

The suggested grid-connected MMI is being tested using a prototype. The prototype of the GCPVS is shown in Figure 5.28. The control technique is carried out with the FPGA controller

Spartan 6E. An emulator Joint Test Action Group (JTAG) is used to dump the control technique. The FPGA

controller's triggering signals have a switching frequency of 1000 Hz. ASS -

P and V - 2SP are the current and voltage sensors implemented to sense the

output current of MMI and the voltage of the grid. The sensors convert sensed current and voltage into low power signals by means of a hall effect sensor.

Such low power signals are fed back to the FPGA controller using A to D (analog to digital) converter.

The THD of the prototype is observed in the power analyzer and the hardware outputs are observed in DSO. Figure 5.29 displays the hardware results of the suggested inverter. Figure 5.30 presents the experimental output of the grid voltage and current of the proposed GCPVS.It is observed that the grid voltage and grid current are in phase during the stable condition and the power factor is near to unity. Figure 5.31 displays the experimental THD of the

grid current. The obtained THD is 2.03% which meets the IEEE519

requirements. The hardware outputs corresponding to the transient condition of the GCPVS is shown in Figure 5.32. The suggested inverter while operating in inductive load provides a active power with a requirement for reactive power

of + 500VAR. The grid current, therefore, starts to lag by 4S,with the grid

voltage. While operating in capacitive load, the Q is -500 VAR and the current in the grid leads the grid voltage by 4S. The suggested inverter, therefore,

pumps sinusoidal current and keeps the voltage of the grid in phase by keeping a unity power factor. The comparison of findings with other methodologies is indicated in Table 5.8.

Figure 5.28 Prototype model of the suggested system

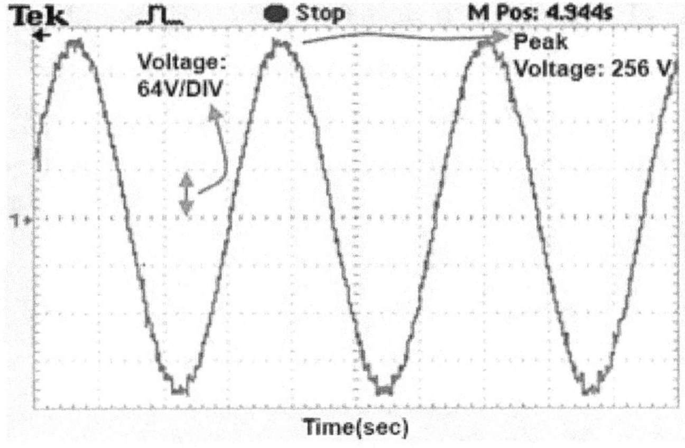

(a)

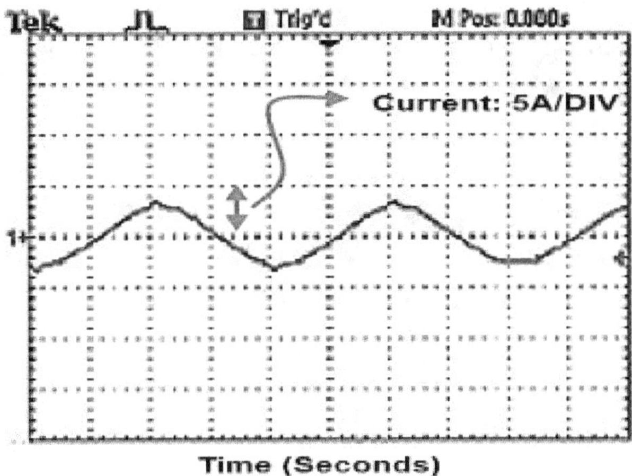

(b)

Figure 5.29 Voltage and Current waveforms of the 31-Level MMI(experimental)

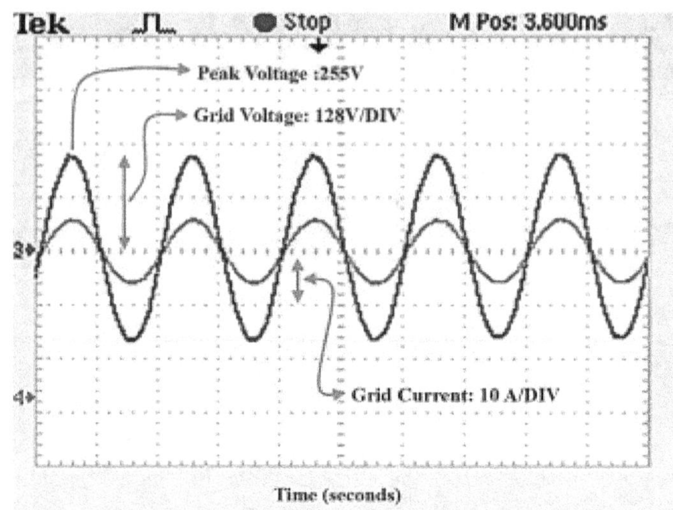

Figure 5.30 Grid Voltage and Grid Current of the Prototype Under Steady State Condition

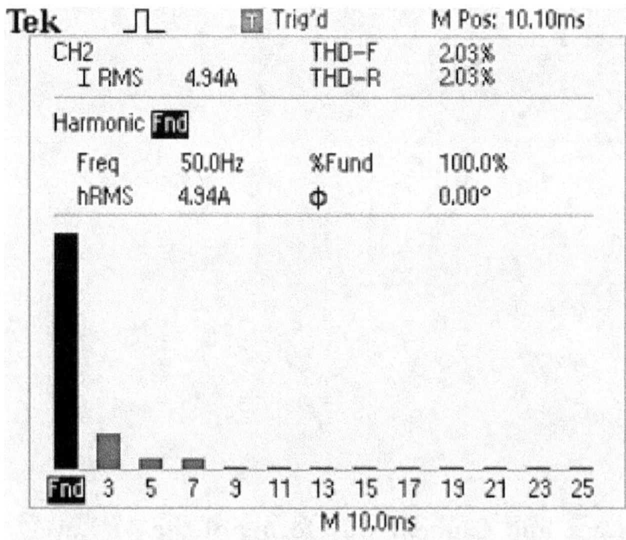

Figure 5.31 Experimental THD of Grid Current

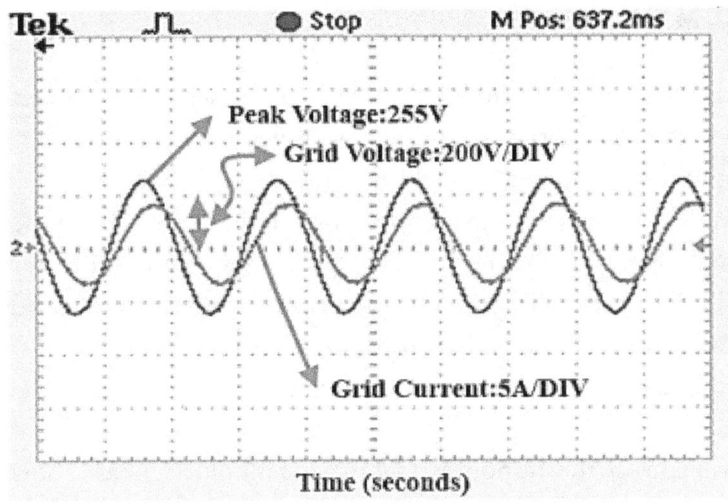

Figure 5.32 Experimental Output of Grid Voltage and Grid Current Under Transient Condition

Table 5.8 Comparison of results with other methodologies

Authors	Techniques	Level	THD
	EANFIS	27	0.78%
Bihari & Sadhu (2020)	PI	27	18.19
	PID	27	16.11
Bana *et al.* (2019)	PI	3	2.6%
Makhamreh *et al.* (2019)	MPC	7	3.45%
Han *et al.* (2019)	SRF-PI	19	11.12%
Satti, & Hasan (2019)	DMPC	7	2.7%
Shi *et al.* (2017)	FFGVF	9	3.54%
	PSO-PI	27	5.97%
Proposed	FF-PI	27	2.47%
	HHO-PI	27	1.49%

5.14 SUMMARY

PV fed MMI is connected to the grid. A closed loop control system with optimization techniques is implemented to introduce harmonic less power to the grid. The P & O technique is implemented to supply constant power from the PV panels to the MMI. The findings obtained indicate that the suggested inverter and HHO-PI controller provide good control when the demand in reactive power varies. The exploration ability in the HHO method aims at generating new and varied responses, while the exploitation ability is mainly responsible for the local improvement of the responses from the exploration phase. This method provides a balance in between the two components. Another key factor that leads to the superiority of this algorithm is the fact that HHO algorithm is able to update the search agents, which eventually lead to finding the global optimum. Therefore, the HHO method produces better results when compared with other methods. The findings slightly differ for simulation and experimental setup. It is because the simulation test is performed under optimal conditions but in the experimental setup the switches suffer certain conduction losses, sensor failures and little disturbances in the implementation of the FPGA controller.

CHAPTER 6

SUMMARY AND SUGGESTIONS FOR FURTHER RESEARCH

SUMMARY

The major research findings are described as follows:

In this research work, the design and development of Modular Multilevel Inverter with improved performance parameters have been carried out with three different topologies such as switches reduction, harmonic elimination, and grid synchronization. The overall objective is to enhance availability, reliability, and power quality using the proposed Modular Multilevel Inverter. The challenges of the PV system and MLI with respect to the implementation of an efficient MPPT controller, increased switch count, switching frequency, and quality of power were completely addressed. The modeling of the solar PV system by considering the variations in the irradiance and temperature has been done which serves as the input source for the MMI.

In switch reduction, the suggested MMI can produce output voltages from 9-level to 31-level depending on the DC inputs. Only 14 switches are implemented to obtain the respective voltage levels depending on the DC inputs. The major benefit of the suggested configuration is that; it can produce the voltage levels without the H-bridge circuit . The proposed structure provides a high value of level to switch ratio.

The output voltage of MLI contains unwanted harmonics. To suppress the harmonics produced by the multilevel inverter many modulation schemes were introduced by the researchers in the past few years. The implementation of a high switching frequency modulation technique for MLI will damage the switches by producing a thermal loss. Therefore, the low switching frequency modulation technique is mostly preferred for harmonic elimination in MLI.

In low switching frequency modulation technique, the selective harmonic eliminating pulse-width modulation (SHEPWM) suits best for removing the harmonic distortions from the inverter output voltage. In this technique, the firing angles for the switches of MLI are obtained by solving the complex trigonometric equations. The proposed work uses three different evolutionary methods namely Particle Swarm Optimization (PSO), Genetic Algorithm (GA), and Bee Colony Optimization (BCO). These methods use an objective function which includes the complex trigonometric equations of the fundamental and low-order harmonics. Moreover, these techniques minimize the objective function to obtain the firing angles which reduce the harmonic distortions.

For integrating PV system to grid a two-stage conversion process is implemented. In the first stage, DC-DC boost converters are implemented for boosting the low voltage obtained from PV panels to a higher voltage and in the second stage, the boosted DC power is converted to AC power using MMI. In the suggested work a closed-loop control system is applied for the GCPVS. The gains of the PI controllers are tuned to inject harmonic less power into the grid. The THD obtained for the PSO-PI controller is 5.97%, the FFA-PI controller is 2.47%, and HHO-PI controller is 1.49%. The derived outputs show that the HHO-PI controller works better than other controllers.

The experimental investigations for the HHO-PI controller is carried out for a 1kWp solar PV system in which the results satisfy the standards for harmonic reduction. The advantages of the proposed method include simple control, less complexity, lower THD, filters, output transformers, and front-end rectifiers.

Table 6.1 Comparison of results for a solar fed modular multilevel inverter

S.No.	Proposed Topology	THD (%)	
		Simulation	Hardware
1.	Selective Harmonic Elimination-BCO	10.43%	-
2.	Selective Harmonic Elimination-GA	7.11%	-
3.	Selective Harmonic Elimination-PSO	5.15%	8.44%
5.	9-level MMI with 14 switches-NR	7.57%	-
6.	27-level MMI with 14 switches-NR	3.62%	-
7.	PSO-Proportional Integral Controller	5.97%	-
8.	FFA-Proportional Integral Controller	2.47%	-
9.	HHO-Proportional Integral Controller	1.49%	2.03%

Table 6.2 Comparison of the results with the existing inverter topologies

Authors	Techniques	Level	THD
Modulation Techniques			
Sandeep *et al.* (2019)	PD	9	15.68%
Sathik *et al.* (2017)	PD	9	13.66%
Mahato *et al.* (2019)	SHE-PWM	9	13.68%
Saeedian *et al.* (2017)	PD	9	11.83%

Table 6.2 (Continued)

Authors	Techniques	Level	THD
Optimization Techniques			
Haghdar .(2020)	TLBO	15	1.45%
	SHE-PSO	7	19.89%
Panda *et al.* (2019)	SHE-FSO	7	14.66%
	SHE-PSFSO	7	10.47%
Kar *et al.*(2019)	FA	11	7.51%
Yong Ning *et al.*(2019)	GA-PSO	3	6.32%
Panda & Panda *et al.*(2018)	PSO	7	11.81%
Memon *et al.*(2018)	APSO	7	12.52%
Grid controllers for Grid Synchronization			
	EANFIS	27	0.78%
Bihari *et al.* (2020)	PI	27	18.19
	PID	27	16.11
Padmanaban *et al.*(2019)	PI	3	2.6%
Makhamreh *et al.*(2019)	MPC	7	3.45%
Han *et al.*(2019)	SRF-PI	19	11.12%
Satti, & Hasan (2019)	DMPC	7	2.7%
Shi *et al.*(2017)	FFGVF	9	3.54%

SUGGESTIONS FOR FURTHER RESEARCH

To extend the research work presented in this thesis, future work may be considered in the following possibilities:

- The suggested method can be extended to the three-phase systems.

- The scheme may be applied to a wind energy conversion system for power quality improvement.

- By extending the concepts and procedures deduced in the thesis, the fault-tolerant capability of the MLI can be carried out.

- The proposed multilevel inverter topology can be applied for various stand-alone medium and high-voltage applications such as induction motor drive, FACTS devices, and HVDC applications.

www.ingramcontent.com/pod-product-compliance
Lightning Source LLC
Chambersburg PA
CBHW070742160726
48004CB00001B/9